WILLIAMSVILLE AND THE WAR OF 1812
REVISED & EXPANDED EDITION

WILLIAMSVILLE AND THE WAR OF 1812
REVISED & EXPANDED EDITION

"GA-DA-YA-DEH" A PLACE OF MISERY

J. A. BUSCAGLIA

A P
ANTIQUARIAN PRESS

Printed in the United States of America

ISBN: 978-1-7370531-5-6

Library of Congress Control Number: 2026903109

Antiquarian Press

P.O. Box 693

East Amherst, New York 14051

www.antiquarianpress.com

For my Father

Charles A. Buscaglia

1917 ~ 1976

"Truth shall spring out of the earth"

— PSALMS 85: 11.

CONTENTS

FOREWORD

On November 4th, 2018, Frank Minnolera from Ground Penetrating Radar company Terracon, expert metal detectorist Jim Bucki and his skilled friends, and I descended on Garrison Park in Williamsville. We were graciously assisted by then Village of Williamsville mayor Dan Delano.

We searched as a team for artifacts from the allegedly historic site, said to be the location of the US Army field hospital during the War of 1812. We did uncover an apparent uniform button (see photo 1.), possible syringe and ball shot (see photo 2.) at the site, but believed many more relics should have been found given the thousands of people who had supposedly spent months of misery during war-time.

Shortly after the article about our quest appeared in the Amherst Bee, two things happened. First, some professional archaeologist-types were upset that we mere amateurs were marring their turf, though they had apparently not considered such an undertaking themselves.

Secondly and most significantly, Joe Buscaglia, author of this tome you are about to read, reached out to me. He explained and methodically proved to my logical satisfaction that our work at Garrison Park had been misplaced.

Having reviewed Mr. Buscaglia's evidence, enclosed herein, which I found quite persuasive, I had to agree we had the right Ta-Ta but the wrong Ho-Ho. It is my belief that Mr. Buscaglia's thesis in this book is absolutely correct.

The site in which we had been looking was the wrong one, marked by an inaccurate plaque.

Mr. Buscaglia, in this book, conclusively demonstrates that by looking in Garrison Park, we were off by less than a mile from the true site, and he has identified the correct location of the sufferings of our forces and people during very bleak times on the Niagara Frontier.

Mr. Buscaglia, Mr. Minnolera, Mr. Bucki and I are currently engaged in projects to responsibly bring to light artifacts associated with not only the cantonment at Williamsville, but personalities associated closely with the founding of the Village of Williamsville, NY, recently listed as one of the USA's best places to live.

Mr. Buscaglia does not hold a doctorate. None of us amateurs have such prestigious pedigrees.

But his evidence-based approach speaks for itself, and Mr. Buscaglia should be commended as a Western New York history treasure.

Our troops, persevering civilians, and the historical record, deserve to be properly honored by correctly identifying the main scene of the wretched times endured by our founders and predecessors on this land.

Let the book you are about to read begin that overdue recognition.

Tom Bauerle

SUNY Buffalo 87

Academic All American

Phi Beta Kappa

History, Communication

Williamsville, NY

2022

Photo 1. Uniform Button

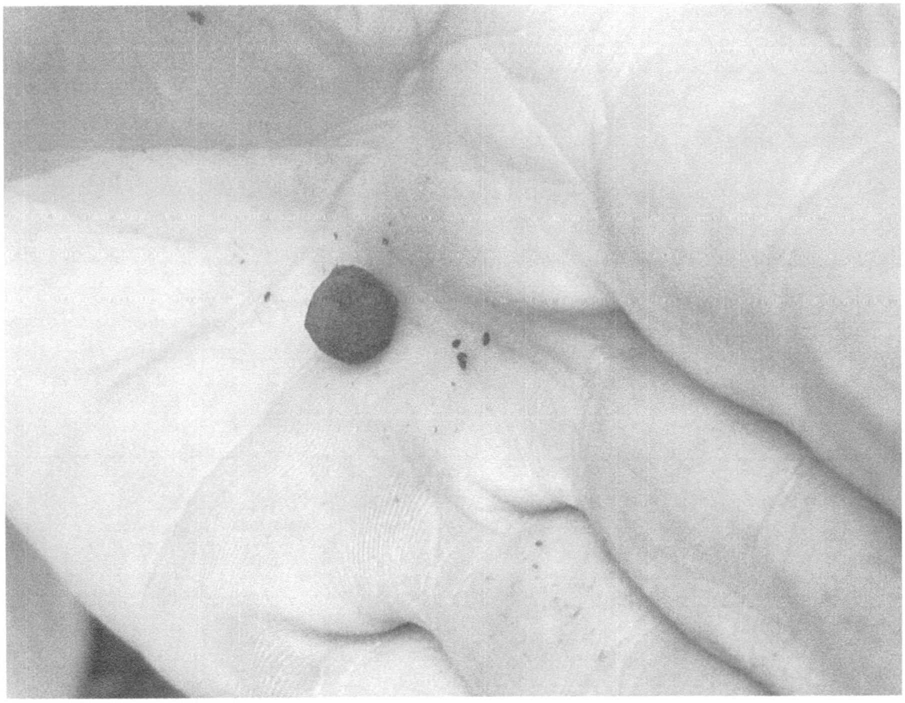

Photo 2. Ball Shot from Garrison Park

INTRODUCTION

During the War of 1812, an army of 5,000 to 6,000 men were quartered at Williamsville, New York. At the time, the Evans House became the temporary headquarters of Generals Smyth, Brown, and Scott.

Long before the settlers came, the Seneca's called the site of present-day Williamsville, New York "Ga-da-ya-deh" which means a place of misery. Maybe they foresaw what was to come, or maybe something had happened in their past?

One thing is certain: between the years of 1812 and 1815, there was plenty of misery at Williamsville, from the executions of deserters to the unimaginable suffering in the general hospital, which accommodated from 300 to 400 sick and a somewhat larger number of wounded.

The only locations which are certain of the Army's encampment, are the War of 1812 Cemetery and The Evans house.

The Seneca's may have seen the future when they named this place.

The diversity of opinion on the locations of the War of 1812, its Barracks and Hospital, being so great that I have written my thoughts and observations on the topic. Hence, I am revealing all the information regarding this subject that I have been able to gather from any source in order to determine the sites. This is simply my opinion from my historical research and observations. It is not an academic paper, but a first-person account of my findings.

J. A. Buscaglia

PREFACE

The pursuit of local history is often sparked by a single moment of discovery. For me, that moment arrived on November 7, 2018, upon reading an article in the *Amherst Bee* titled, "Artifacts Discovered at Garrison Park: Digging Up the Past". While I was encouraged to see the Village of Williamsville actively investigating its early 19th-century roots, my own thirty-five years of research into the region's War of 1812 history lead me to a different conclusion. I do not believe Garrison Park is the site of the historic War of 1812 Hospital ruins.

To understand why, we must look beyond the surface and examine the historical record of land ownership, topography, and contemporary military accounts.

The Question of Land and Ownership

During the War of 1812, the land now occupied by Garrison Park was owned by Isaac F. Bowman. Crucially, there are no records indicating that Bowman ever leased this land to the United States government. The chain of title further complicates the "Garrison" narrative:

• Following the war, Bowman lost the land in an 1819 sheriff sale to Joseph Ellicott.

• Upon Ellicott's death in 1826, the property passed to his brother-in-law, Lewis E. Evans.

• While the army purportedly used the Evans house (then located at the corner of Main and Miller, now Main Street and Oakgrove Drive) as a temporary headquarters, this does not equate to the presence of a permanent hospital installation on the park grounds.

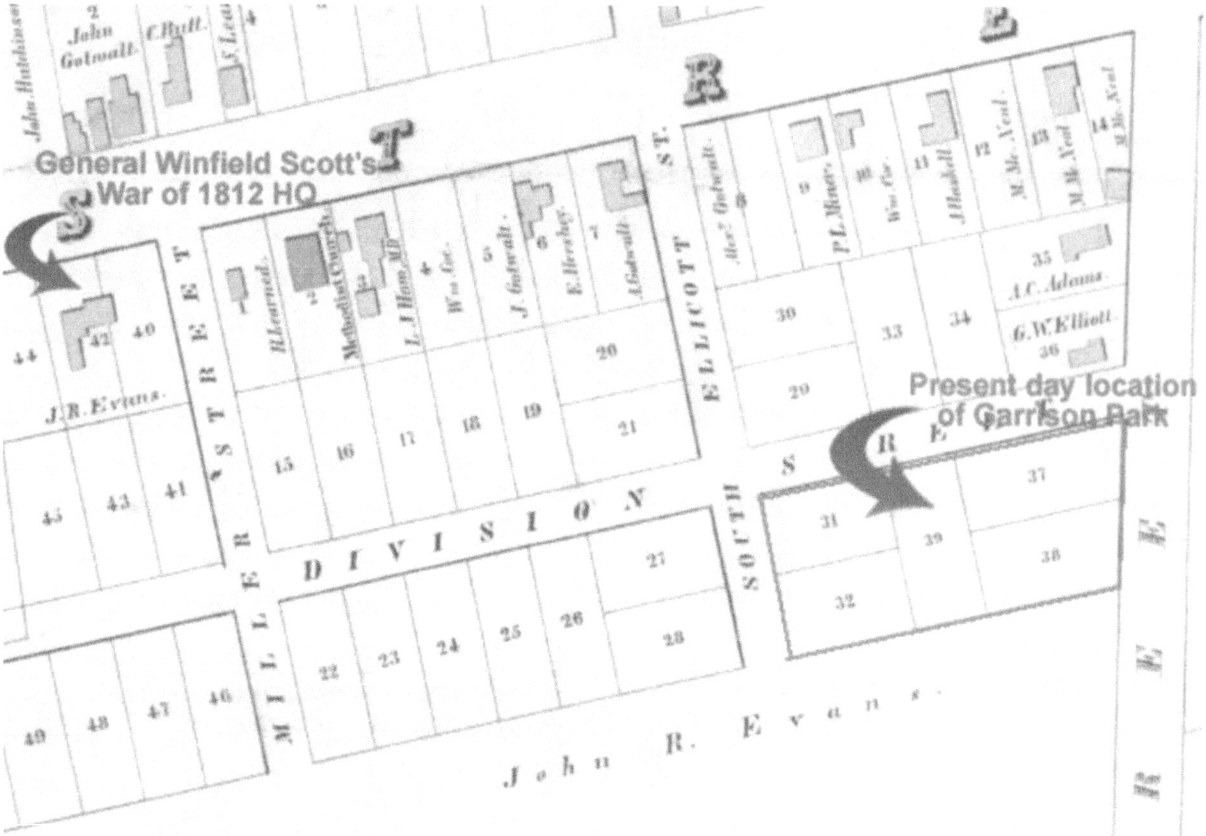

Figure 1 1854 Map of present day Garrison Park property owned by John R. Evans

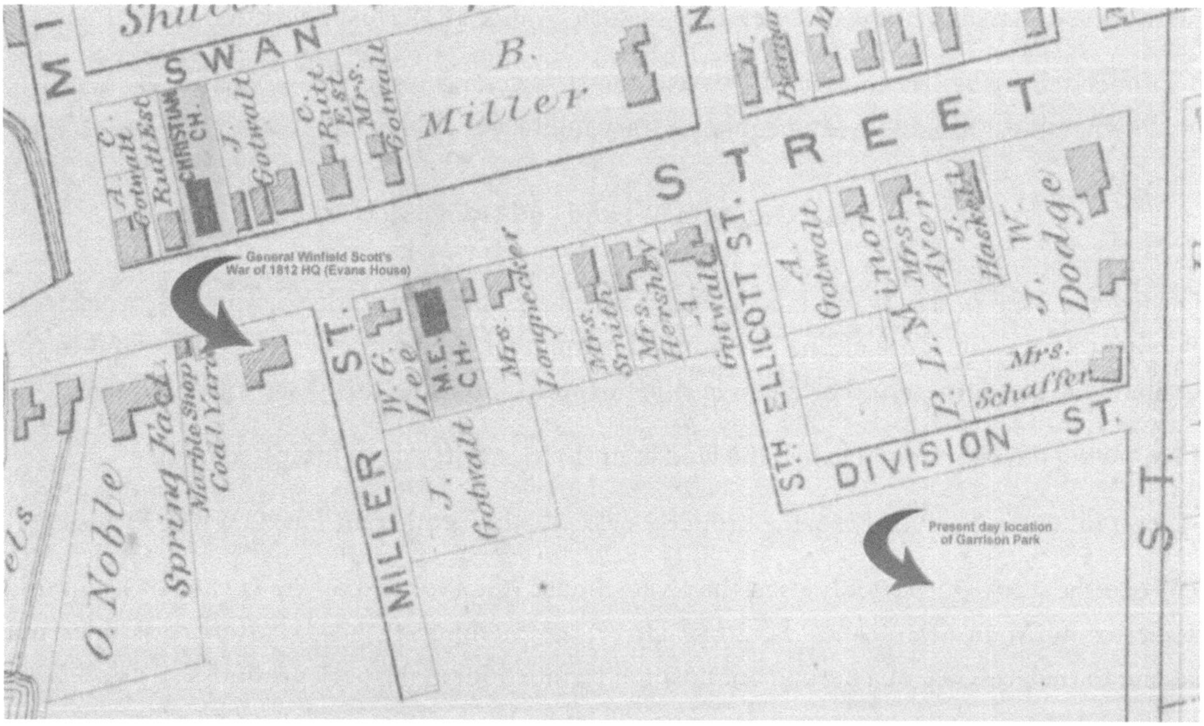

Figure 2 1880 Map of present day Garrison Park area

Environmental and Practical Constraints

The physical state of the land in the early 1800s would have made it a poor choice for a military cantonment. Historical records, including the 1917 *Documents of the Assembly of the State of New York*, describe the area as a marsh containing two large springs and a creek that frequently flooded.

"At the point where the springs emerge from the ground a basin was excavated, walled with rubble and covered with a timber roof... it may have been due to infiltration into the reservoir from a small stream which crosses this park, particularly at times when this stream was in flood."

Given these conditions, it is highly improbable that the army would have selected a flood-prone marsh for a hospital or barracks. Furthermore, because the site later served as the Village Reservoir, any artifacts discovered there in recent years were likely deposited as fill during efforts to reclaim the marsh for the construction of the reservoir and standpipe. (see Figure 4)

218 CONSERVATION COMMISSION

Williamsville is a municipality in the town of Amherst, Erie county. It was incorporated as a village in 1869 and still remains such an incorporated village. The water supply affairs of this village are administered by a separate Board of Water Commissioners. Williamsville is situated on Ellicott creek and the Niagara Falls branch of the Lehigh Valley railroad, about four miles northwest from Depew. It is connected with Buffalo by a trolley line.

Since 1895 a municipally owned water supply system has been operated in this village. Water was obtained from two large springs in the bottom of a reservoir, located in a park in the southeasterly portion of the village, which park is bounded on the north and east by Division and Garrison streets. At the point where the springs emerge from the ground a basin was excavated, walled with rubble and covered with a timber roof. This basin is 50 feet by 30 feet in plan and 6 feet deep. Water ordinarily stood in it to a depth of 5 feet, giving a storage of 55,000 gallons. From this basin the water was conducted by gravity to South Ellicott street, thence to Main street and to the power house of the Buffalo and Williamsville Railroad Company, situated on the south side of Main street and on the right bank of Ellicott creek. In this power house, two pumps, said to have a rated capacity of 750,000 gallons per day, force the water into the distribution system. These pumps are run by power furnished by the railroad company. A portion of the water is carried in a southeasterly direction to the corner of Miller and Division streets and up Division street to a steel standpipe situated in the southeasterly corner of Division and South Ellicott streets within the same park which contains the reservoir. This standpipe is 12.5 feet in diameter, 100 feet high and holds 80,000 gallons when full. Cast-iron distribution piping has been laid in the built up streets of the village and the necessary fire hydrants have been connected to it.

The State Department of Health has reported from time to time that the water supplied to the people of Williamsville from this reservoir was frequently polluted to a dangerous extent. The source of this pollution is not clear. It may have been due to the seepage of sewage from the nearer buildings into the limestone caverns through which this spring water undoubtedly passes, or it may have been due to infiltration into the reservoir from a small stream which crosses this park, particularly at times when this stream was in flood.

For some years the yield of this spring proved sufficient to supply the people of Williamsville with water, but during the dry season of 1915 the supply suddenly failed entirely and at the time of the hearing had not come back to any extent. In this emergency the village authorities were obliged to take drastic measures to maintain a supply of water in the village and to that end the four wells above mentioned were sunk.

On June 28, 1915, the Board of Water Commissioners of the village of Williamsville petitioned the Trustees of that village for authority to spend the sum of $2500 for the sinking of these wells. By resolution, duly adopted at a special meeting held that same day, the Board of Trustees of this village authorized the said Board of Water Commissioners to borrow that amount from some bank in Buffalo. This was accordingly done and the work proceeded.

Figure 3: Documents of the Assembly of the State of New York, *Volume 22, 1917*

(see Figure 4)

Figure 4 Williamsville from Erie County 1909, New York Published by Century Map Co. in 1909

Contemporary Evidence and Proximity

The most persuasive evidence, however, lies in the testimony of those who were there. In *Snake Hill: An Investigation of a Military Cemetery from the War of 1812*, the authors note that in July 1813, a site for a new general hospital was selected based on the recommendation of Dr. Ezekiel W. Bull. This hospital was built on 90 acres leased from Raphael Cook not Isaac Bowman.

Furthermore, primary sources and firsthand accounts consistently place the hospital approximately one mile upstream along the creek. Garrison Park simply does not fit this geographic description.

The Scope of This Work

The following chapters outline my theory regarding the true location of the Hospital Cantonment, the associated huts, the Rifle and Artillery Cantonments, and the Arsenal. I will also identify the site of the War of 1812 Cemetery, which, while originally part of the Town of Amherst, now falls within the boundaries of the Town of Cheektowaga.

Finally, I must extend my deepest gratitude to Mr. Joseph Whitehorne. His generosity in sharing his extensive research and notes from the National Archives regarding Raphael Cook was invaluable in piecing together this historical puzzle.

—J. A. Buscaglia

1
HUT'S AND HOSPITAL'S

Growing up near the War of 1812 cemetery, I was always aware of its quiet presence and the sense of mystery that surrounded it. As a child, I remember passing the cemetery for the first time and being captivated by the old cannon and the rows of white crosses standing in solemn formation. Those early impressions stayed with me, and as I grew older, my curiosity expanded beyond the cemetery itself to the landscape around it. I recall my first hike along Ellicott Creek from Youngs Road at Aero to Maple Road and the first time I paddled down the creek in a cheap vinyl raft, letting the current carry me through a place steeped in history. These experiences formed the earliest threads of a lifelong fascination.

One memory in particular stands out: seeing the historical plaque on Main Street at Garrison Road in the Village of Williamsville. The sign reads:

U.S. Barracks 1812 *Along Garrison Road to creek and extending southeast were log barracks of Gen. Smyth's army during winter of 1812 and '13. These buildings were later used as hospital.*

Erected State Education Department–1939

At first, the sign puzzled me. Why would military planners build their barracks away from Ellicott Creek? The inscription suggests that the barracks extended "along Garrison Road to creek and extending southeast," which would lead one away from the water source. And who would willingly carry water any farther than necessary, especially during a harsh frontier winter? That question became the spark that ignited my long search to understand the true location and nature of the Williamsville Barracks.

Over time, the Barracks and Hospital in Williamsville have taken on an almost mythical quality. They are often mentioned, speculated about, and imagined, yet never definitively located. In many ways, they have become the "Unicorn of the Village" a place everyone believes existed, but no one has ever truly seen.

One of the first discoveries in my research was that the site was not referred to as "barracks" or a "garrison" in contemporary documents. Its official name was the Cantonment at Williamsville, and the structures were called huts. This raised an important question: what exactly was a cantonment?

During the American Revolution, both British and American forces typically suspended active campaigning during the winter months. The landscape became difficult to traverse, and essential resources especially food and forage were scarce. In the eighteenth century, quartering troops during military campaigns generally took one of three forms: encampment, billeting, and cantonment.

Encampments consisted of tents arranged by regiment and brigade according to the line of battle. They were temporary, easily assembled and disassembled, and designed for mobility. However, tents were wholly inadequate for the severe winters of the northern frontier.

Billeting involved quartering troops in existing houses and public buildings. This method required the least labor and resources, but it was deeply unpopular among civilians. The resentment toward billeting was so strong that the Founders addressed it directly in the Third Amendment. Nevertheless, billeting still occurred on the Niagara Frontier. For example, the *Buffalo Gazette* of 10 January 1815 reported:

The house of Mr. Wm. Long of Williams' Ville, occupied by the U. States troops, was accidentally consumed by fire on the night of the 7th instant.

Cantonments, by contrast, were a middle ground. They consisted of semi-permanent structures built in an organized fashion to house soldiers during winter or extended pauses in campaigning. They offered greater stability than tents and avoided the political and social tensions associated with billeting.

In short, a cantonment was a more permanent encampment used for winter quarters.

My first step in unraveling the mystery was to locate the oldest maps of the Village of Williamsville and the Town of Amherst. The earliest I found were from 1854 and 1866, both showing Garrison Road. Prior to that name, the road was known as Lancaster Road as late as 1851, and earlier still as Colborn Settlement Road, according to a deed I located.

My second step was to research the War of 1812 and the settlement known today as Williamsville. Many modern accounts claim that during the war the area was known as "Williams Mills." This is not supported by contemporary documents. In letters written by military officers and doctors stationed there, the authors consistently referred to the place as Williamsville or Williams Ville. Likewise, the site was never called "the Barracks" in official correspondence—it was always the Williamsville Cantonment.

From *Death at Snake Hill: Secrets From a War of 1812 Cemetery* by Joseph W. A. Whitehorne and colleagues:

The hospital at Williamsville, about 11 miles east of Buffalo. This site was picked on 29 July on the recommendation of the senior surgeon, Dr. Ezekiel W. Bull. Thus, 90 acres and the stables of Raphael Cook's farm were leased for the construction of a general hospital...

The book describes the mass graves, the segregation of British and American dead, and the rapid construction of a tent city followed by more permanent brick structures. Captain John Larkin, the supervisory quartermaster, brought in skilled workers from as far away as Rochester and Utica. Locally produced brick was hauled from Buffalo to Williamsville, and glass and shingles were purchased to complete the buildings. Patients began arriving on 30 July.

Multiple sources—including *Snake Hill: An Investigation of a Military Cemetery From the War of 1812*—state that 37 hectares (approximately 91.5 acres) of Raphael Cook's farm were leased for the hospital. Cook certainly existed; he died in Buffalo in 1822. However, I have found no deeds showing he owned land outside the Village of Buffalo.

> Raphael Cook, (1755-1822), served in Capt. Nodiah Hooker's company, 1776, that occupied Boston after the evacuation. He was born at Southington, Conn., and died at Buffalo, N. Y.

Lineage Book by Daughters of the American Revolution

William Ketchum's *An Authentic and Comprehensive History of Buffalo* (1864) provides additional insight into Raphael Cook, describing him as a tavern keeper whose establishment—later known as the Old Phoenix Hotel stood for many years on Main Street. Ketchum notes that Cook may have arrived in Buffalo around 1810, operating a modest public house that hosted community meetings and travelers. After the burning of Buffalo in December 1813, Cook returned and reopened his tavern on the site where the Tifft House would later stand. His presence in the region during the War of 1812 is further supported by a death notice published in the *Buffalo Gazette* on January 24, 1815, which reads:

Died in Wm's Ville, on Tuesday last, Mrs. Anna Cook, consort of Capt. Raphael Cook, aged 54.

> *DIED*—At Colchester, Ct. July, 13, 1813, dea. *Noah Skinner*, aged 88: on the 10th ult. Mrs. *Sarah Skinner*, his wife, aged 88. This couple were married in the year 1745, consequently lived as husband and wife 66 years.
>
> In Canandaigua, Mr. [illegible] Richards, formerly of the firm of Norton & Richards.
>
> ⇨ In Wm's. Ville, on Tuesday last, Mrs. *Anna Cook*, consort of Capt. Raphael Cook, aged 54.

January 24, 1815, in the Buffalo Gazette Death Notice Anna Cook

This notice confirms that the Cook family was living in or near Williamsville during the period when the hospital was constructed. Yet despite Cook's documented presence, I have found no evidence that he owned land outside the Village of Buffalo. This raises an important question: if Cook did not own the land, whose property was leased for the hospital?

To answer this, it is necessary to consider the essential requirements for constructing a large military hospital or barracks in the early nineteenth century. The most critical need was fresh water. Creek water was unreliable and often unsafe, especially for sick or wounded soldiers. A dependable spring was far more desirable. This is where Lehn's Springs becomes significant. The springs provided abundant, clean water and were located outside the village limits an ideal location for a cantonment.

The land surrounding Lehn's Springs was owned by Gamaliel St. John, who purchased several lots in Township 11 Range 7 and Township 12 Range 7. According to Holland Land Company survey maps, Township 11 Range 7 is technically part of Cheektowaga, not Amherst. St. John's holdings included what is now the Lehn Springs development. In 1807, he sold the farm to John Frick, who moved there with his wife and seven children after emigrating from Lancaster County, Pennsylvania.

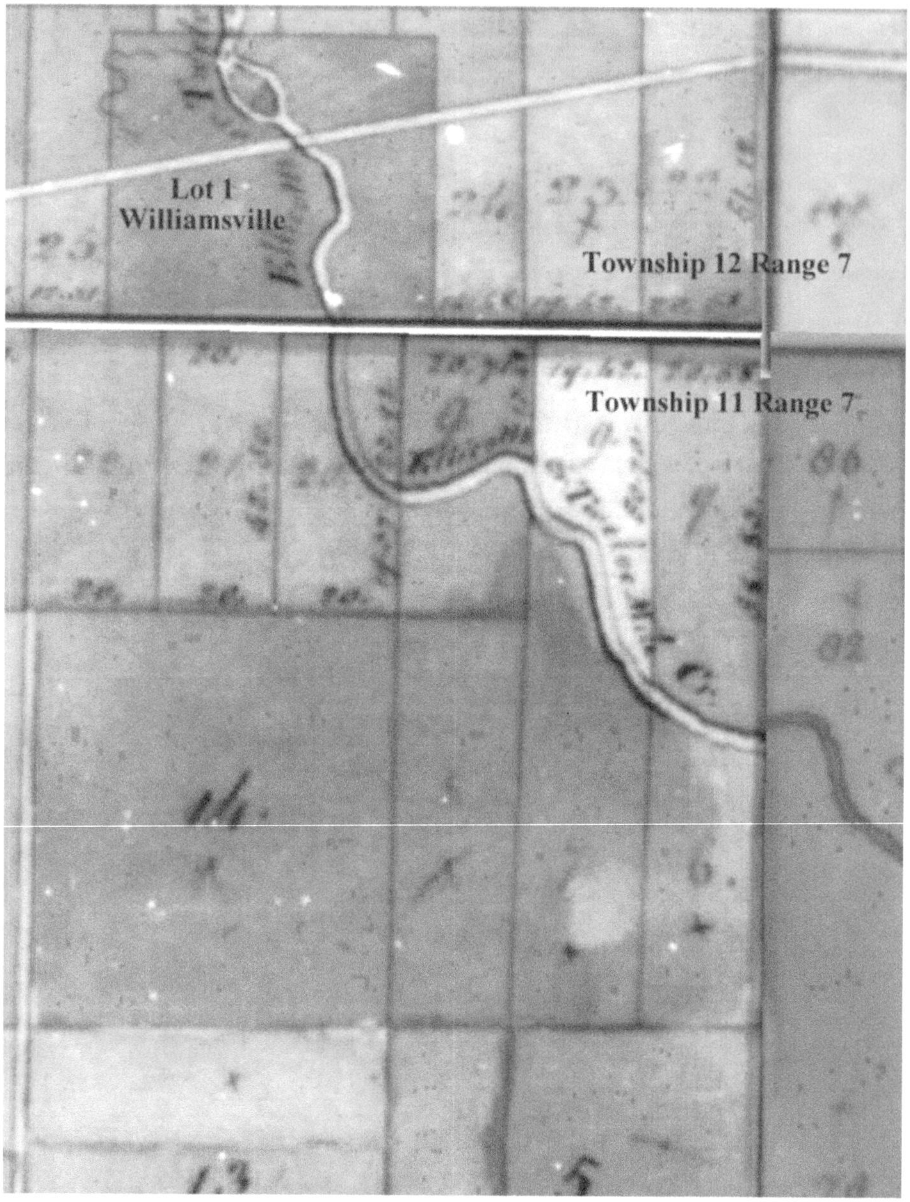

Holland Company Survey Maps of Townships 11 & 12 Range 7

The outline is the boundary of Gamaliel St. John's farm purchased in 1807 shown on a Holland Land Company Map.

Gamaliel St. John sold the farm to John Frick, most recently known as the Lehn Farm. The Frick family lived there until the War of 1812.

A 1923 article in the *Amherst Bee*, written by Ida L. Zent and titled "History of the Lehn Family," provides a detailed account of the Frick family's early years in the region. The article states:

"The Lehn family were among the early settlers in Williamsville. John Frick... purchased a farm of Mr.

St. John, now known as the Lehn farm. The family lived there until the War of 1812. The barracks, as we all know, are about one-fourth of a mile from this farm."

This statement is remarkable for its matter-of-fact tone. In 1923, the location of the barracks was apparently common knowledge so well known that the author felt no need to elaborate. The article also recounts that John Frick died while caring for sick and wounded soldiers and was initially buried on his farm before later reinterment at Harris Hill.

After the burning of Buffalo in 1813, the Frick family returned to Pennsylvania. It is plausible that during their absence, Raphael Cook who had fled Buffalo after the burning—leased the Frick property and then sub-leased part of it to the government for the construction of the hospital. This theory is supported by a lease agreement preserved in the National Archives (RG94 OAG QM Accounts, John B. Hogan, Box 90), which states:

"At Doctor Bull's recommendation, on 29 July, 90 acres and the stables of Raphael Cook's farm in Williamsville were leased for construction of a general hospital. Rent was $500 every three months."

The contract also included a notable stipulation: "no burying place be suffered on the premises." This may reflect Cook's awareness that he did not own the land and wished to avoid creating permanent alterations that might upset the Frick family upon their return.

The reference to "90 acres" is particularly significant. Lots 9 and 20 of Township 11 Range 7 both part of the Frick farm—are each approximately 90 acres. This alignment between the lease description and the size of the Frick property strengthens the case that the hospital was built on land belonging to the Frick family, not Cook.

The Frick/Lehn cabin stood near the northeast corner of present-day Garrison Road and Wehrle Drive. From this location, a quarter-mile radius extends into Lots 9 and 20, precisely where the cantonment would logically have been situated. This area also aligns with the presence of Lehn's Springs, providing the fresh water necessary for a large military installation.

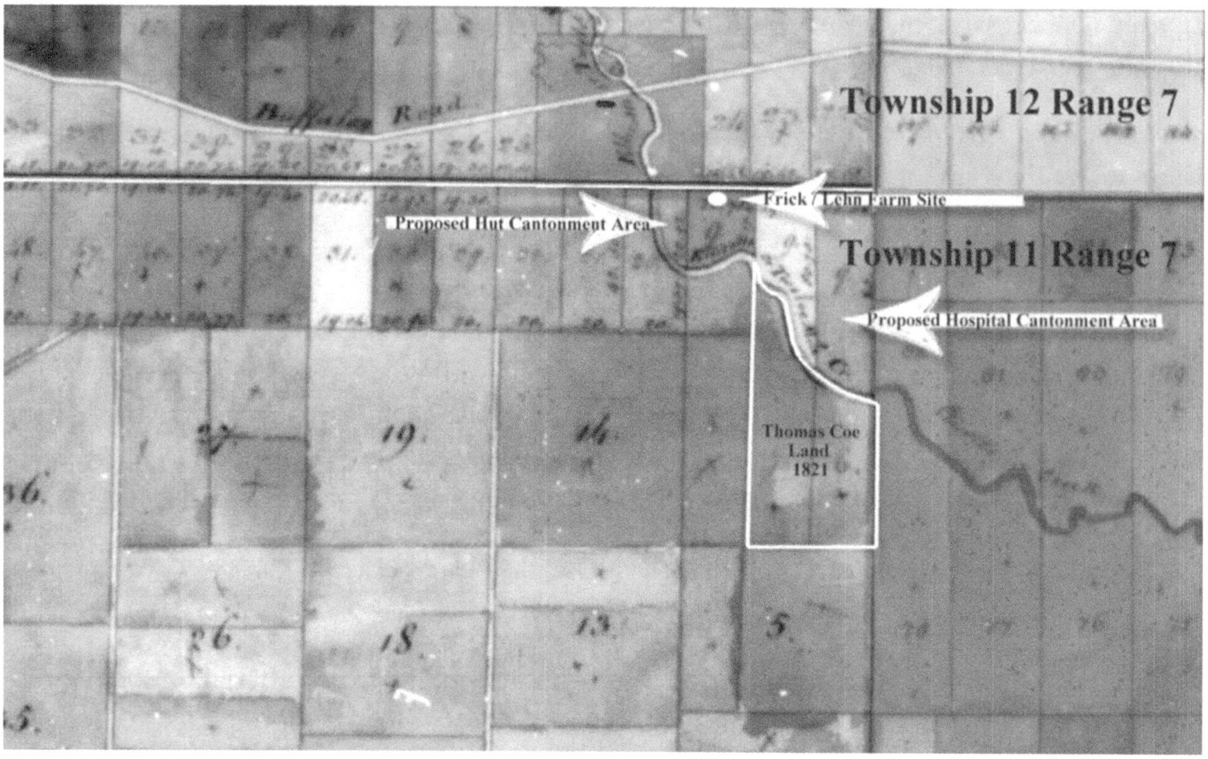

Holland Company Survey Maps of Township 11 & 12 Range 7 with the Proposed Search Area

The 1/4-mile radius from the Frick/Lehn Farm would also extend into Lot 9 & 20 of Township 11 Range 7 (See the above Holland Land Company Survey Maps). The circle is the Frick/Lehn Farm site. It was near the northeast corner of present-day Garrison Road and Wehrle Drive.

I have also outlined the Thomas Coe Land boundaries on the map based on two references I found to the Coe family and the Cantonment.

First, I found that Thomas Coe purchased Lots 6 & 7 in Township 11 Range 7 containing 202 acres on October 17th, 1821, recorded in Library 7, Page 207.

Further evidence comes from the Coe family. Thomas Coe purchased Lots 6 and 7 in Township 11 Range 7 in 1821. Two obituaries recorded in Janet Wethy Foley's *Early Settlers of New York State* provide compelling clues.

Coe, Charlotte. w of Thomas D., native of Fulham, near London, Eng. Jan. 4, 1833, 36y. D at the cantonment near Williamsville.

Coe, John A., Dec. 31, 1819, 28y.

Coe, Thomas, May 21, 1838, 80y. D at the cantonment near Williamsville.

Obituaries in the Book, Early Settlers of New York State

The first reads: *"Coe, Charlotte, wife of Thomas D., native of Fulham, near London, Eng. Jan. 4, 1833, 36y. Died at the cantonment near Williamsville."*

The second states:

"Coe, Thomas, May 21, 1838, 80y. Died at the cantonment near Williamsville."

These entries are extraordinary. They indicate that the cantonment was still recognized as a physical location decades after the War of 1812. Moreover, the wording "near Williamsville" suggests that the cantonment was not located within the village itself but in the surrounding countryside consistent with the Frick/Lehn farm area.

Sue Miller Young's *A History of the Town of Amherst, New York* adds further support. She notes that the area remained a military station for many years after the war, with records of a "very wonderful New Year's ball" held there as late as 1848. She also cites an advertisement placed in the *Buffalo Gazette* on August 14, 1814, by Dr. E. W. Bull, Hospital Surgeon, offering a reward for a stolen horse evidence of ongoing military activity at the site.

40 *Dollars Reward.*

TAKEN from the Hospital Lot, near Wms. Ville, on the night of the 20th inst. a sorrel HORSE, about 14 hands high, nick'd and bobb'd, with two white hind feet, and an unclosed wound in his chest. The horse is an excellent gig, but an indifferent saddle horse. Whoever will secure and deliver to the subscriber, the horse and thief, shall receive 40 dollars reward, or a liberal compensation for the horse alone.

N. B. Likewise, a *STUD HORSE*, 4 ys. old, dark bay, natural trotter, the end of his tail a little squared. about 15 1-2 hands high. The same reward will be given as is above mentioned.

E. W. BULL,
Hospital Surgeon.

Wms. Ville, Aug. 29, 1814. 3w46

From the Buffalo Gazette, *August 29, 1814*

The Cantonment wasn't in Williamsville! Ok, so they have plenty of fresh water at Amherst/Lehn Spring for a Cantonment and it's not in Williamsville but near, and it is ¼ mile from the Frick/Lehn farm.

So, on the following pageI found this Aerial Photo of the area from 1927.

Aerial Photo of the Area from Fairchild Aerial Surveys 1927

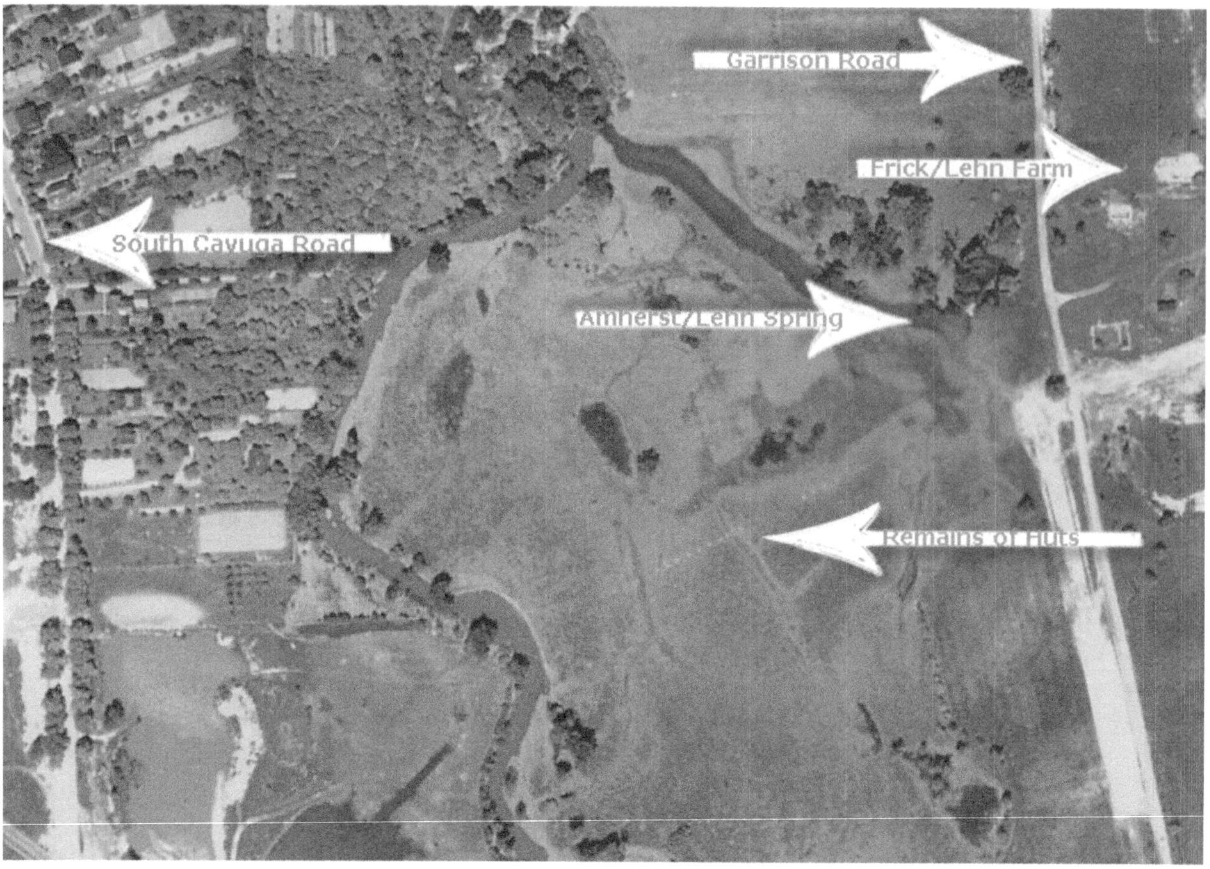

Aerial Photo with Site Markers from Fairchild Aerial Surveys 1927

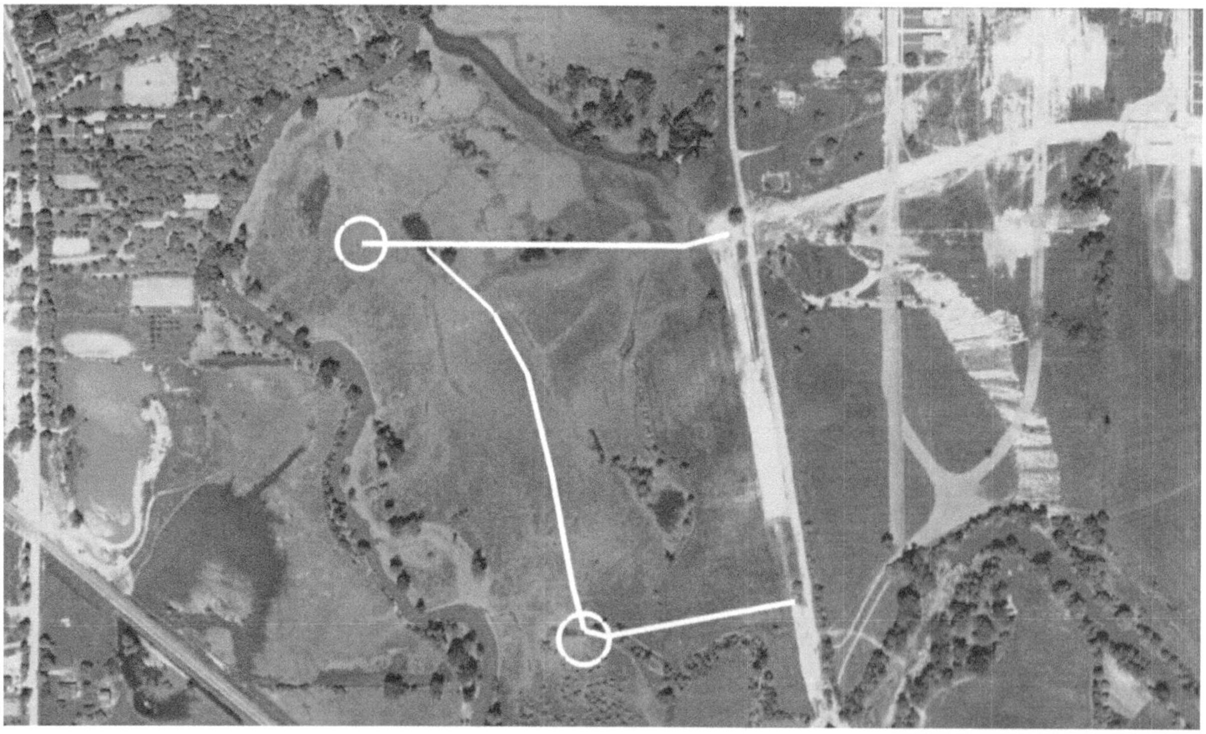

Aerial Photo of the Frick/Lehn Farm Area from Fairchild Aerial Surveys 1927 with road drawn in place.

The final piece of evidence emerged from a 1927 aerial photograph taken by Fairchild Aerial Surveys. When I examined the image of the area surrounding the Frick/Lehn farm, I noticed a line of evenly spaced white dots in the middle of an open meadow. At first, their purpose was unclear. They appeared too regular to be natural features, yet too isolated to be part of a modern structure.

After considerable study, the answer became clear: these white dots were the stone remains of the fireplaces used in the soldiers' huts.

Their alignment, spacing, and location correspond perfectly with what one would expect from a large cantonment constructed during the War of 1812. The site lies outside the historic village limits, in what is now the Lehn Springs development precisely where the historical evidence points. The alignment of these stone features in the 1927 aerial photograph is striking. They appear in a straight, deliberate line precisely the kind of arrangement one would expect from a military cantonment constructed with standardized hut spacing. Their presence in an otherwise open meadow, with no modern structures nearby at the time, strongly suggests that they are remnants of the War of 1812 encampment. The fact that they appear in the exact area indicated by historical accounts near the Frick/Lehn farm, within a quarter-mile radius, and adjacent to a reliable freshwater source further reinforces this conclusion.

To understand the significance of these remains, it is helpful to consider how military huts of the period were constructed. Soldiers typically built small log structures with stone fireplaces at one end. These fireplaces were essential for heating, cooking, and drying clothing during the harsh winter months. When the huts were abandoned or dismantled, the wooden components would decay or be removed, but the stone fireplaces often remained as durable markers of the former encampment. Over time, these stones would collapse into small piles precisely the kind of features visible in the aerial photograph.

The discovery of these stone alignments provides a rare physical link to the Williamsville Cantonment. For decades, the site had been discussed in historical accounts but never definitively located. The combination of documentary evidence, land records, family histories, and aerial photography now offers a coherent and compelling picture of where the cantonment once stood.

The broader historical context further supports this interpretation. During the War of 1812, the Niagara Frontier was a volatile and strategically important region. The American army required a secure location to house troops, treat the wounded, and prepare for military operations. Williamsville, situated along a major transportation route and near abundant natural resources, was an ideal location. The presence of Lehn's Springs provided a reliable source of fresh water, while the surrounding farmland offered space for constructing huts, hospitals, and supply depots.

Moreover, the cantonment's location outside the village proper aligns with military practices of the time. Armies often established winter quarters on the outskirts of settlements to avoid overcrowding, reduce the risk of disease transmission, and maintain a degree of separation between soldiers and civilians. The Frick/Lehn farm area, with its open fields and proximity to water, would have been well suited to these needs.

The historical record also indicates that the cantonment was a substantial installation. The rapid construction of a tent city, followed by more permanent brick structures, suggests a significant military presence. Captain John Larkin's efforts to bring in skilled workers from Rochester and Utica demonstrate the scale and urgency of the project. The purchase of large quantities of brick, glass, and shingles indicates that the hospital buildings were intended to be durable and functional, capable of housing large numbers of patients.

The presence of mass graves at the War of 1812 cemetery further underscores the importance of the Williamsville Cantonment. The cemetery contains the remains of both American and British soldiers, buried in separate sections a practice consistent with military customs of the time. The proximity of the cemetery to the cantonment suggests that it served as the primary burial ground for those who died at the hospital.

The lease agreement specifying that "no burying place be suffered on the premises" is particularly noteworthy. This clause likely reflects the landowner's desire to prevent permanent alterations to the property. However, the existence of mass graves nearby indicates that the need for burial space eventually outweighed contractual restrictions. The cemetery's location just outside the leased area would have satisfied the terms of the agreement while still providing a practical solution for the burial of the dead.

The story of the Williamsville Cantonment is also intertwined with the experiences of local families. The Frick family's involvement in caring for sick and wounded soldiers, and John Frick's death from overwork, highlight the human toll of the war on the civilian population. The Lehn family's long tenure on the property, and their recollections of the barracks' location, provide valuable insights into the site's history. The Coe family's obituaries, noting deaths at the cantonment decades after the war, suggest that the site continued to be recognized as a distinct location well into the nineteenth century.

Taken together, these pieces of evidence form a cohesive narrative. The Williamsville Cantonment was located near the Frick/Lehn farm, within a quarter-mile radius of the farmhouse, and adjacent to Lehn's Springs. The stone remains visible in the 1927 aerial photograph are likely the remnants of the soldiers' hut fireplaces. The site served as a major military installation during the War of 1812, housing troops, treating the wounded, and supporting operations along the Niagara Frontier.

The process of uncovering this history has been both challenging and rewarding. The lack of surviving structures, the passage of time, and the development of the surrounding area have obscured the physical traces of the cantonment. Yet through careful research, analysis of historical documents, and examination of aerial imagery, it is possible to reconstruct the story of this important site.

The Williamsville Cantonment may no longer be visible on the landscape, but its legacy endures in the historical record, in the memories of local families, and in the quiet presence of the War of 1812 cemetery. The discovery of the stone remains in the aerial photograph provides a tangible connection to the past, offering a glimpse into the lives of the soldiers who once lived and died there.

2
THE HUT'S

I have referenced the first soldier huts built in the area that I found below Eleven Mile Creek which is now called Ellicott Creek; at the time it was called Eleven Mile Creek because the point where it crosses Main Street is approximately 11 miles to the Village of Buffalo.

Lieut. Patrick McDonogh to his Parents.

WILLIAMSVILLE, December 12th, 1812.

. . . .

. . . .

. . We are now encamped in the woods, building huts which we expect to get into by the middle of next month. It is rather late in the season to be in tents. We have a very handsome situation on Eleven Mile creek. The place is called after its owner, a Colonel Williams of New York. I hear he contemplated building his house next spring on the very ground on which we are building, and desired that not a piece of timber should be cut, as he wished it entirely shaded, but I can promise him that by that time there will not be a sapling standing within a mile of it. We marched from Black Rock to this place on the 11th inst. You have heard ere this of the duel that took place on the 12th between Generals Smyth and Porter, and the latter's statement of facts. General S. has left here on furlough. Col. Porter of the Light Artillery commands in his absence, and as we are eleven miles from the enemy everything is quiet.

(From advance sheets of the Records of the American Catholic Historical Society of Philadelphia.)

Letter dated December 12, 1812 Lt. Patrick McDonogh to his parents

The book, *The Early Firm of Juba Storrs & Co.,* Rev. Albert Bigelow 1874, also states that they established the Barracks and the Hospital's location to be about a mile from Williamsville up the creek (See Figure 14.)

> The declaration of war at once brought business to the young firm, especially at Williamsville. They made a contract with the Government for all the mill products they could furnish, for the army, and had work enough to do. There were no other mills nearer than Niagara Falls, except, as Lucius Storrs with somewhat of poetic license was wont to say, one that wasn't bigger than a coffee mill. Then, in the winter of 1812-13, barracks and a hospital were established about a mile from Williamsville up the creek, by cutting down trees and building huts in the woods.

Passage from Rev. Albert Bigelow, The Early Firm of Juba Storrs & Co. *p. 117*

They also mention the Hospital Barracks in the U.S. Army Medical Department Office of Medical History:

> The hospital at Williamsville was located in an extensive barracks which had been renovated for use as a hospital. Since some of the men brought to this facility lacked adequate clothing, it was doubly fortunate that it was possible here to keep the rooms warm. Only six men were assigned to each ward, the rooms were easy to keep clean, and the patients tended to recover their health quickly. Having settled his patients in Williamsville's wholesome accommodations, Mann set out to join General Wilkinson at his winter camp at Malone, leaving surgeon's mate Joshua Whitridge, a physician whose "services . . . cannot be too highly appreciated," in charge. It seems probable that this facility, too, was closed at some time in the winter of 1813–14 or the spring of 1814.[57]

The Army Medical Dept, I 775 - I 8 I 8 Early Campaigns in the North: 1812-1813 p.169

Footnote #57 for the last passage:

57. Secretary of War to Major General Hampton, 25 Sep 1813, Cruikshank, *Documents*, 2: 107; Mann, *Sketches*, pp. 95, 96–97, 117, quote from pp. 96-97. Joseph Lovell's report of the summer of 1814 states that a hospital was *established* at Williamsville on 1 August 1814, implying that for some time before that date, there was no hospital in operation there (Lovell report, Mann, *Sketches*, p. 162).

The Army Medical Dept., I 775 - I 8 I 8, Early Campaigns in the North: 1812-1813 p.260

I believe they constructed the same Soldier Huts in the War of 1812 as they constructed during the Revolutionary War at Valley Forge. I don't believe Hut Technology advanced much during that short period of time.

Valley Forge Soldiers Hut model.

There is very little eyewitness evidence for the building of these huts and only the briefest contemporary documentation which shows each hut was to be built by the twelve men intended to occupy it. Details on the intended hut design are contained in Washington's Orderly Book:

There is only one (to my knowledge) eyewitness account of the building of the winter huts at Valley Forge. It is derived from the diary and recollections of soldier Joseph Plumb Martin and was published in the 19th century as *A Narrative of a Revolutionary Soldier*. The complete text of his description of hut construction is shown below, right, but specific extracts will be inserted where relevant in the following discussion:

The next thing is the erecting of the huts; they were generally about twelve by fifteen- or sixteen-feet square, (all uniformly of the same dimensions,) the building of them was thus; after procuring the most suitable timber for the business, it was laid up by notching them in at the four corners. When arrived at the proper height, about seven feet, the two end sticks which held those that served for plates were made to jut out about a foot from the sides and a straight pole made to rest on them, parallel to the plates; the gable ends were then formed by laying on pieces with straight poles on each, which served for ribs to hold the covering, drawing in gradually to the ridgepole. Now for the covering; this was done by sawing some of the larger trees into cuts about four feet in length, splitting them into bolts, and driving them into shingles, or rather staves; the covering then commenced by laying on those staves, resting the lower ends on the poles by the plates, they were laid on in two thicknesses, carefully breaking joints; these were then bound on by a straight pole with withes, then another double tier with the butts resting on this pole and bound on as before, and so on to the end of the chapter. A chimney was then built at the center of the backside, composed of stone as high as the eves and finished with sticks and clay, if clay was to be had, if not with mud. The last thing was to hew stuff and build us up cabins or births to sleep in, and then the buildings were fitted for the reception of gentlemen soldiers, with all their rich and gay furniture.

3
CONCLUSIONS REGARDING HOSPITAL CANTONMENT & SOLDIERS HUTS

I believe that the cantonment containing both the soldiers' huts and the military hospital was located in what is now the Lehn Springs Development. The evidence documentary, cartographic, geographical, and photographic converges on this area with remarkable consistency. The most compelling physical clue is the line of white dots visible in the 1927 Fairchild aerial photograph. These dots, isolated in the middle of an open meadow, represent the last surviving piles of stones from the fireplaces of the soldiers' huts. When viewed alongside the historical accounts, land records, and military correspondence, they form a coherent picture of the cantonment's true location.

One of the earliest and most valuable accounts comes from Ida L. Zent's paper on the Lehn family, read at the Old Home Day Reunion on August 4, 1923. Zent stated that the family farm lay less than a quarter mile from the barracks and that John Frick died of exhaustion while caring for sick and wounded soldiers. Her matter-of-fact phrasing "as we all know" suggests that the location of the barracks was common knowledge in 1923, even if that knowledge has since faded from public memory.

Additional confirmation appears in the obituaries of Charlotte and Thomas Coe, both recorded in *Early Settlers of New York State*. Each obituary states that the individual "died at the cantonment near Williamsville." This phrasing is significant. It indicates that the cantonment was not located *in* the village, as later generations assumed, but *near* it. The Coe family's land lay close to the area now known as Lehn Springs, placing them in proximity to the hospital cantonment described in the obituaries.

Cartographic evidence further supports this conclusion. The Holland Land Company survey maps show that Lot 20 long associated with the Frick and Lehn families was located in Township 11 Range 7, not Township 12 Range 7. This distinction matters because Township 11 Range 7 corresponds precisely to the area where the aerial photograph reveals the line of stone features. The alignment of historical lot boundaries with the physical remains strengthens the case for this location.

Military correspondence from the War of 1812 adds yet another layer of corroboration. In a letter dated December 12, 1812, Lieutenant Patrick McDonogh wrote to his parents from Williamsville, stating that the troops were "building huts on the Eleven Mile creek." This description aligns perfectly with the geography of the Lehn Springs area, which lies along Ellicott Creek historically known as Eleven Mile Creek. McDonogh's letter provides a contemporary, firsthand account of hut construction in precisely the area under consideration.

Further evidence appears in the *Publications of the Buffalo Historical Society*, Volume 4, in an account titled *The Early Firm of Juba Storrs & Company*, read by Rev. Albert Biglow in 1874 and published in 1877. Biglow places both the hospital and the barracks "about a mile from Williamsville up the creek." When measured along the known wagon roads of the period, the distance from the Evans House then serving as the Army's temporary headquarters to the proposed cantonment site in the Lehn Springs area is approximately 0.87 miles. The route would have followed the Batavia Road (present-day Main Street) eastward, then turned south onto Colborn Settlement Road (present-day Garrison Road), leading directly to the cantonment just west of what is now Lehn Springs Drive. This distance matches Biglow's description with striking precision.

The aerial photograph from 1927 provides the final, tangible piece of evidence. In the image, a line of evenly spaced white dots appears in the middle of an otherwise empty meadow. These dots correspond to the collapsed stone bases of the fireplaces used in the soldiers' huts. Their spacing, alignment, and isolation strongly suggest intentional placement rather than natural formation. When overlaid with Von Steuben's 1802 *Plan for the Order of Encampment of a Regiment Making but One Battalion*, the arrangement of the dots aligns with the expected layout of huts in a standardized military encampment.

Taken together, these clues lead to a single, consistent conclusion:

The cantonment containing the barracks, huts, and hospital of the War of 1812 was located in the present-day Lehn Springs Development.

If resources permitted, this is the area where archaeological investigation should be focused.

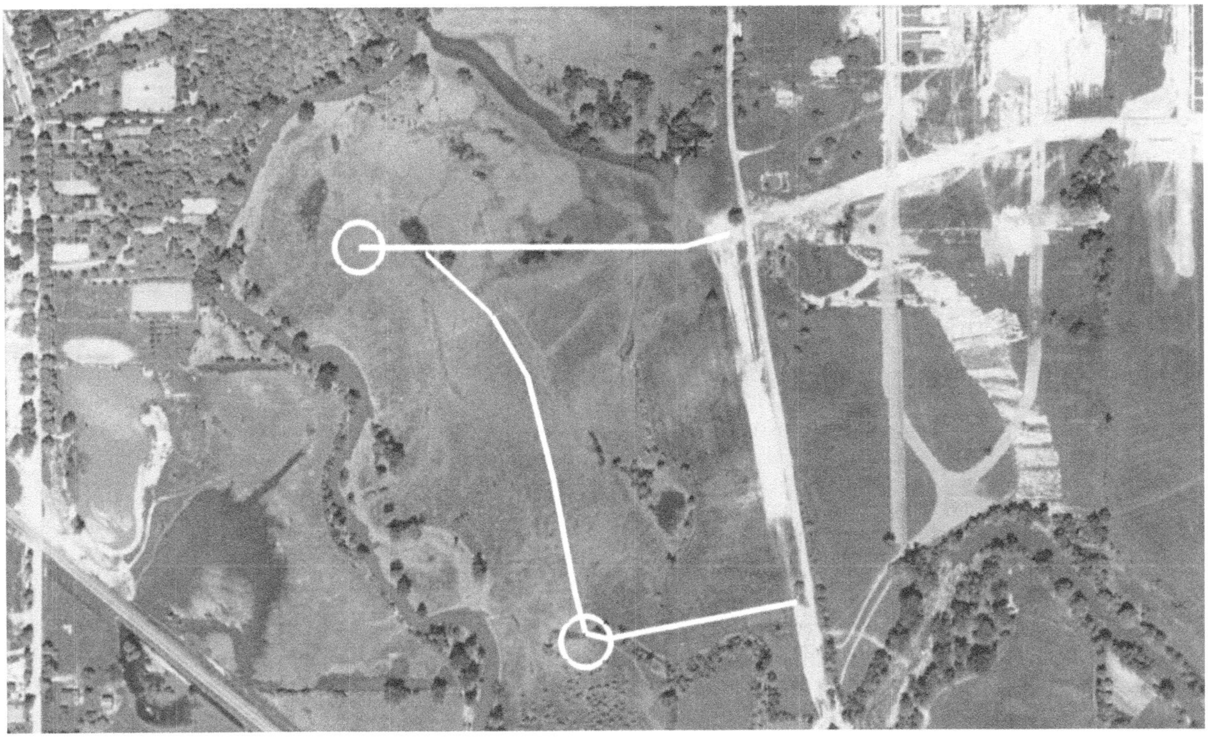

Aerial photograph of the Lehn Springs area from Fairchild Aerial Surveys (1927), showing the line of white dots believed to be the remains of hut fireplace bases.

Von Steuben's "Plan for the Order of Encampment of a Regiment Making but One Battalion" overlaid on the Lehn Springs Development area, illustrating the alignment between the historical encampment layout and the stone features visible in the 1927 aerial photograph.

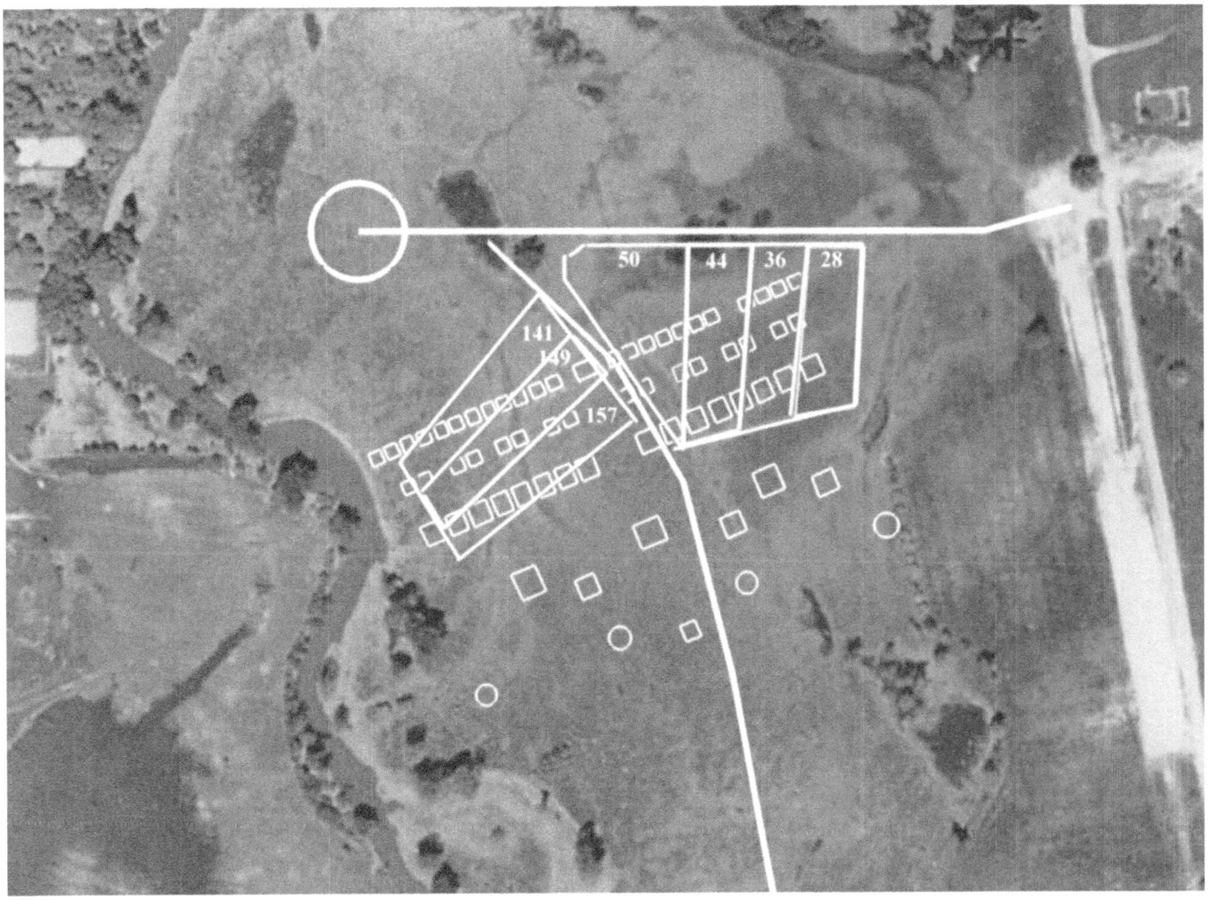

Present-day Lehn Springs Development with proposed hut locations and lot lines. The white squares represent War of 1812 soldiers' huts, each approximately 16' × 14'. Their positions are based on the stone piles visible in the 1927 aerial photograph. Lot lines correspond to current street addresses on Lehn Springs Drive.

The white squares represent the War of 1812 soldiers' huts. Each one is approximately 16' x 14'. The piles of stones from the 1927 Aerial Photo were used to position the locations. The upper first row is where the stone piles were located. The lot lines and numbers represent their present street addresses on Lehn Springs Drive.

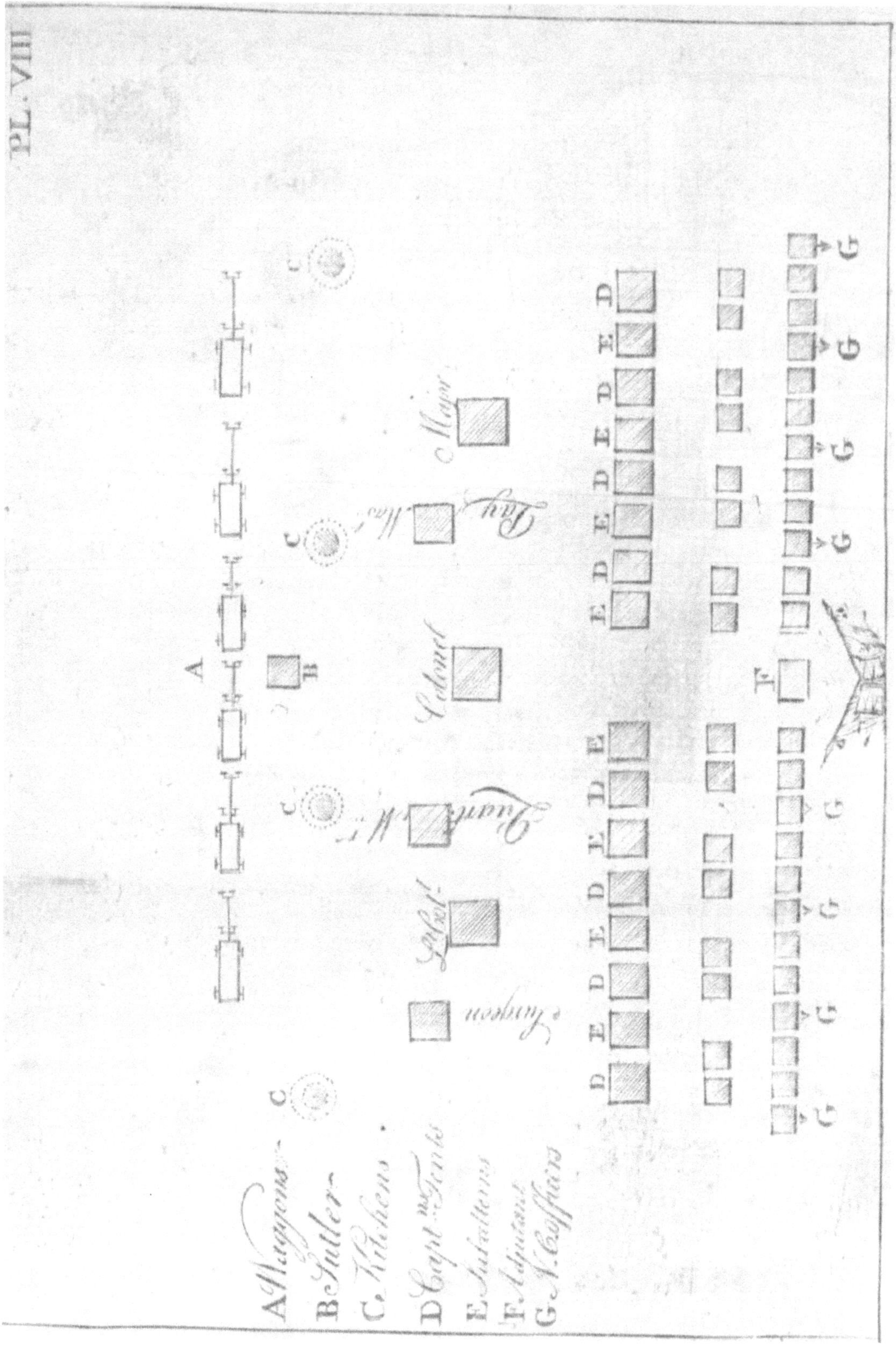

Von Steuben's "Plan for the Order of Encampment of a Regiment Making but One Battalion," Plate VIII, from *Regulations for the Order and Discipline of the Troops of the United States* (1802).

. . .

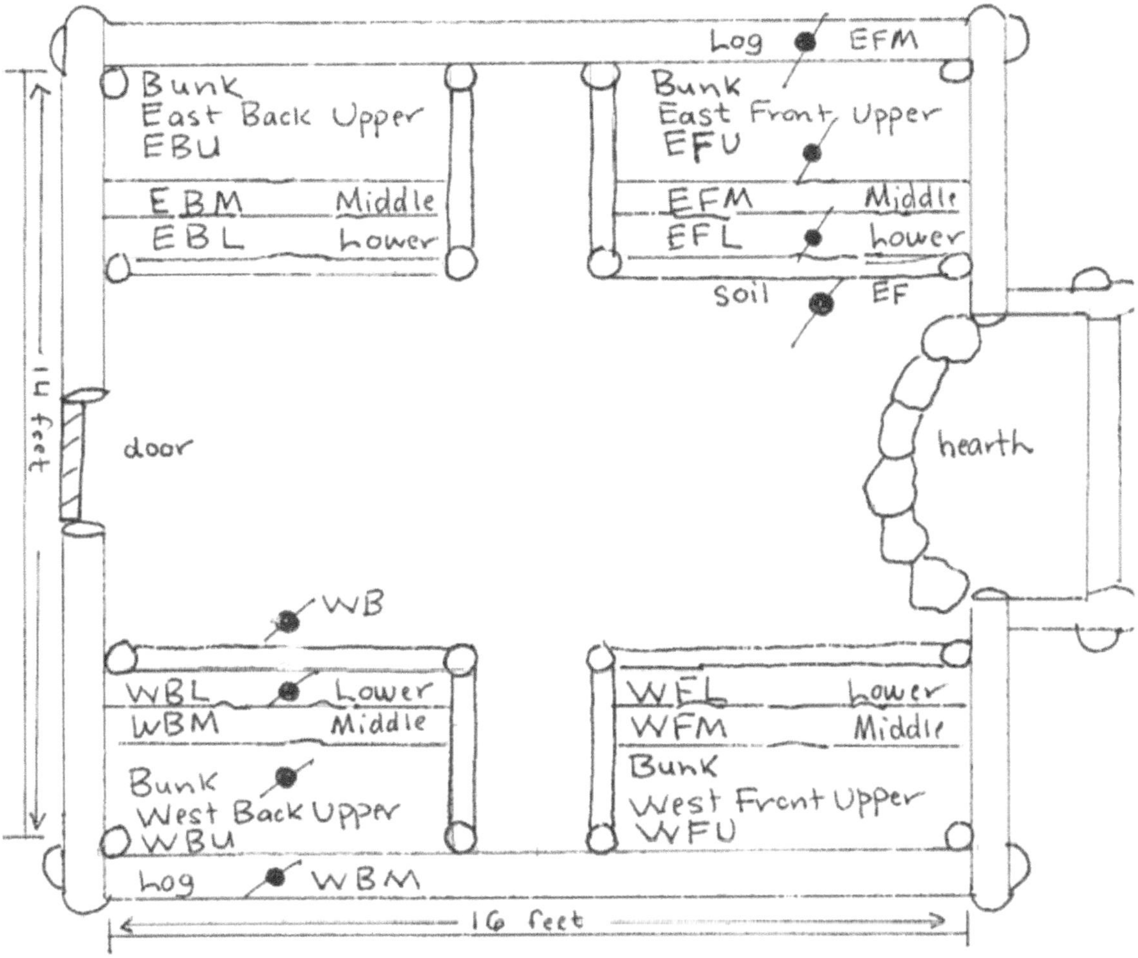

Soldiers Hut Dimensions from Valley Forge Models by Phil Lord

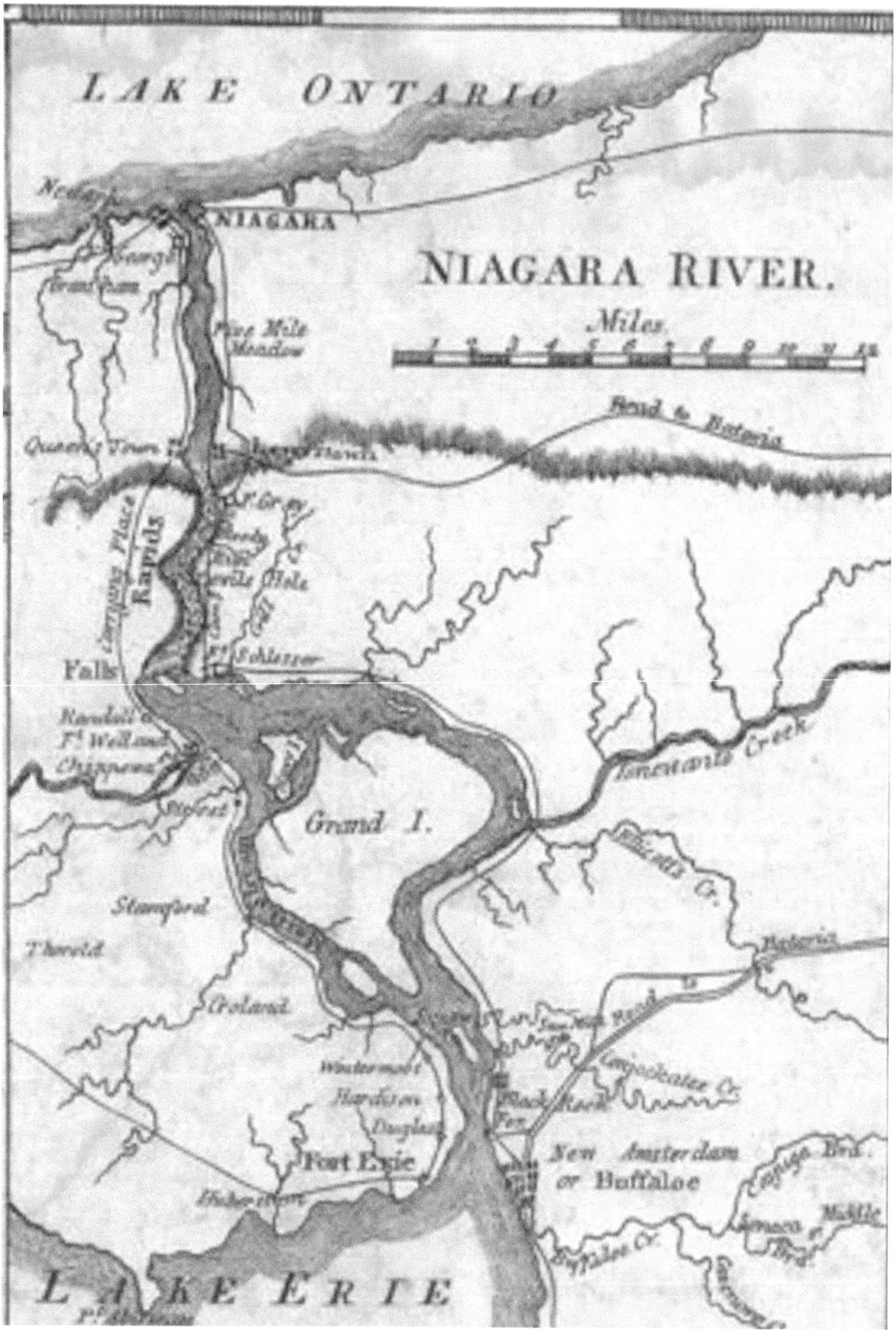

1815 map showing two circles just below the Batavia Road where it crosses Ellicott Creek (present-day Williamsville, NY). These circles likely designate the locations of the U.S. Army Headquarters and the Hospital Cantonment. The distance between the circles is approximately one mile, consistent with contemporary accounts describing the hospital as "one mile up the creek."

4

MAY 7TH, 2023, THE DAY I'VE WAITED 40 YEARS FOR!

On May 7, 2023, I received an email that marked a turning point in a search I had pursued for four decades. The message arrived unexpectedly, yet it addressed precisely the subject that had occupied my research and imagination for so many years: the lost War of 1812 hospital and cantonment.

The email read:

Greetings J.A. Buscaglia:

I have recently located your piece on the internet concerning the War of 1812 hospitals and possible locations.

I purchased the above-mentioned property in 2018. There is a functional fireplace sitting in the center of the backyard that was concealed by hydrangea trees that had been growing the last 50 years.

When I had the trees removed two years ago from the structure, I noticed how old the quarry rock appeared, specifically, on the chimney. My neighbors had invited an architectural historian to have a look at the fireplace and potentially render a date for the age of the stone. He estimated 200+ years in age.

I have contacted the Village of Williamsville about this and have asked former owners of the property and it has been a uniformed response of not knowing anything about the history of the fireplace.

The house was built in 1925 and the land was previously owned by John Lehn. My property is a one-minute walk to the creek.

I am sending you photos for your review, and you are welcomed to stop by to look at the fireplace.

The photographs that accompanied the email revealed a remarkable structure. When first discovered, the fireplace had been completely engulfed by a hydrangea bush more than fifty years old; only the hearth was visible. Once the overgrowth was removed, the full form of the fireplace emerged massive, weathered, and unmistakably old.

The hearth faced away from the house. The back of the fireplace, oriented westward, showed the same rough-cut quarry stone. An architectural historian had already examined the structure and estimated the stonework to be more than two centuries old. The south-facing side measured 45.5 inches in width, while the north-facing side displayed similar proportions. In 2024, the late Town of Cheektowaga Historian, Maureen Gleason, inspected the fireplace as well.

Fireplace as first discovered, largely concealed beneath a hydrangea bush more than fifty years old; only the hearth was visible.

Hearth of the fireplace, facing away from the house.

West-facing rear of the fireplace.

An architectural historian has look at the fireplace and potentially render a date for the age of the stone. He estimated 200+ years in age.

South-facing side of the fireplace, measuring 45.5 inches in width.

North-facing side of the fireplace.

The late Town of Cheektowaga Historian Maureen Gleason inspecting the fireplace, 2024.

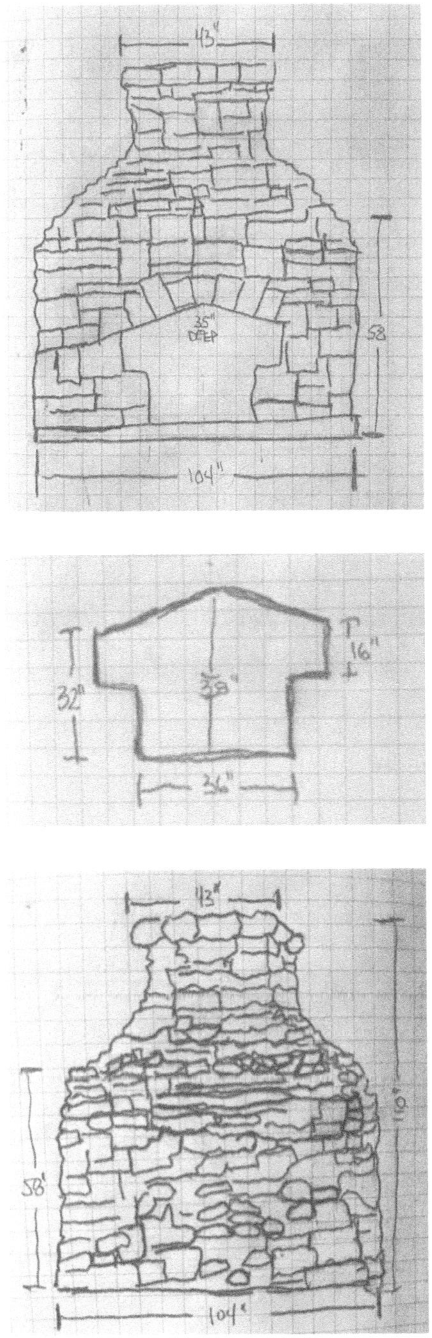

Fireplace dimensions: 45.5-inch thickness, 104-inch face, 110-inch height.

Fireplace Dimensions

The fireplace is monumental in scale. Its thickness measures 45.5 inches; the face spans 104 inches; and the height reaches 110 inches—though it was almost certainly taller when originally constructed. The interior is so large that an adult man can stand upright inside it. This is not the scale of a hut or officer's quarters. It is the scale of a structure designed to heat a substantial building.

More Supporting Evidence

Additional evidence emerged during a sewer line replacement across the street from the fireplace. Construction workers unearthed hundreds of bricks in an area where no bricks should have been present. This discovery aligns with quartermaster reports describing the Williamsville hospital as a brick-built facility.

From *Snake Hill: An Investigation of a Military Cemetery from the War of 1812*:

Captain John Larkin was named supervisory quartermaster for the construction of the Williamsville Hospital. He brought in skilled workmen from as far away as Rochester and Utica.

Huge quantities of locally produced brick were purchased in Buffalo and hauled to Williamsville to be used for construction of the hospital barracks.

The presence of these bricks buried, displaced, and forgotten corresponds directly with the documented construction of the hospital.

Location where hundreds of bricks were unearthed during sewer line replacement, consistent with documented brick construction of the hospital.

Location where hundreds of bricks were unearthed during sewer line replacement, consistent with documented brick construction of the hospital.

1927 Aerial Photo of Huts and Hospital Area

The 1927 aerial photographs of the huts and hospital area reveal extensive bulldozing, suggesting that much of the original site was destroyed during early twentieth-century development. I believe the only reason the fireplace survived is because the first owner of the house, Archie Harkness, recognized its significance and insisted on its preservation.

Harkness was an extraordinary figure in local history. A lifelong educator and community advocate, he had the house built on the property where the fireplace stands. The BOCES Harkness Center on Aero Drive bears his name. He played a key role in establishing Millard Fillmore Suburban Hospital—the second suburban hospital in Williamsville. The first, in a poignant twist of history, stood in his own backyard.

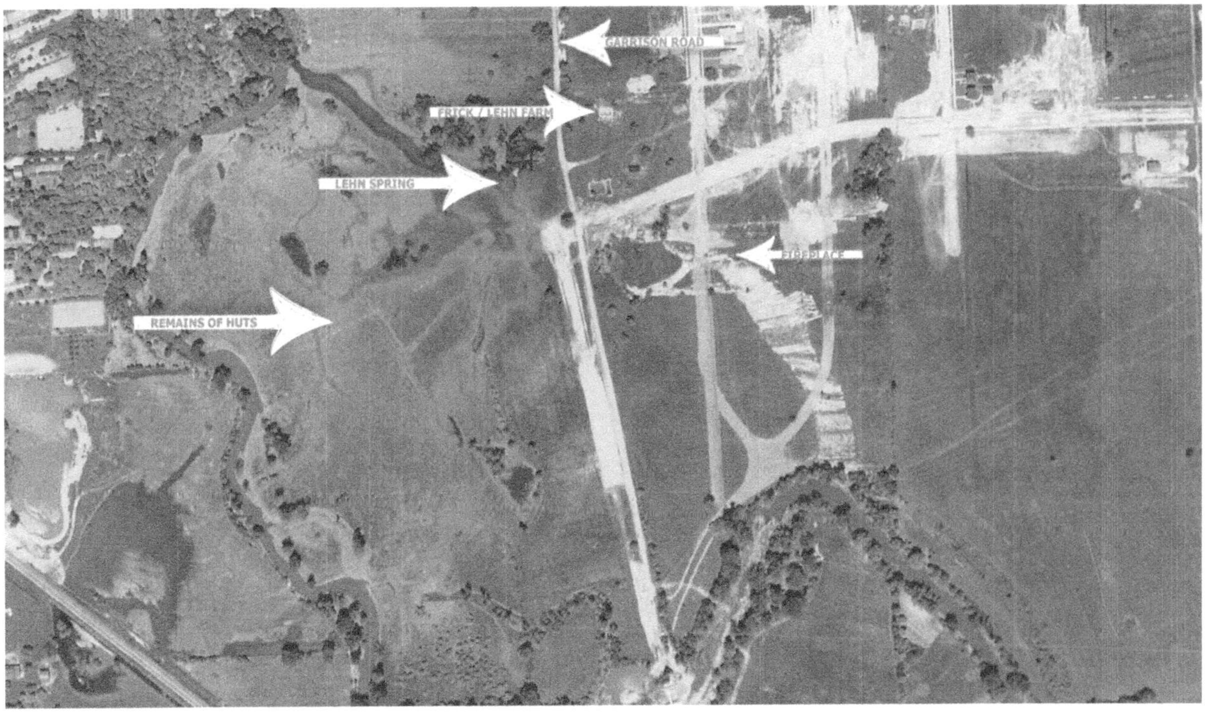

1927 aerial photograph showing extensive bulldozing in the huts and hospital area.

Present day Satellite Imagery of Huts and Hospital area.

Modern satellite imagery of the huts and hospital area confirms the alignment between the fireplace, the creek, and the historical footprint of the cantonment. The fireplace stands today as the last surviving physical remnant of the War of 1812 in this area. Confronting it in person, one can almost sense the sorrow, pain, and suffering witnessed by those stones more than two centuries ago.

I am profoundly grateful to the current property owner for allowing me to visit, examine, and reflect upon this extraordinary artifact. It is a rare and tangible link to a chapter of local and national history that has long been obscured.

Present-day satellite imagery of the huts and hospital area.

5
REFERENCES OF THE WILLIAMSVILLE CANTONMENT IN ORDER, ARTICLES, LETTERS & REPORTS

During the winter of 1812, General Alexander Smyth retired his army to winter quarters in Williamsville, NY. Down the Garrison Road, just south of Lehn Springs, a series of huts were built for the Artillery troops which is where they quartered until the spring of the following year. The huts would later be used as a hospital.

There is some confusion regarding where the regular army wintered in WNY throughout the winter of 1812-13. Although the army remained in Buffalo at Flint Hill until the beginning of December 1812, they then retired to Williamsville. There is no record of any huts being built at Flint Hill and it would have been unlikely that soldiers would be expected to live in tents during a Buffalo winter.

The following confirm the army was quartered in Williamsville.

General Order.

Headquarters, Cantonment Williamsville.
December 8th, 1812.

By a return of the brigade of General Tannehill of Pennsylvania volunteers it appears that five captains, four lieutenants, eleven ensigns, eighty-three sergeants, eighty-nine corporals, twenty- five musicians and nine hundred and thirty privates had revolted and deserted, leaving for duty only two hundred and sixty-seven privates. In consequence thereof General

Tannehill will be pleased to organize the remaining non-commissioned officers, musicians and privates in a battalion under the command of Major Harriett. A furlough for the residue of the time for which he has been called into service is granted to Brigadier-General Tannehill and hi» brigade staff. A furlough for the residue of this month is granted to the field officers of General Tannehill's brigade, except Major Harriett; from who expiration of which furlough they shall be considered as dis- charged the service of the United States.

By order,

Harold Smyth, Lieutenant 3rd Artillery, A. D. C.

(American State Papers.)

Brigadier-General Smyth to Governor Tompkins.
Williamsville, 11th December 1812.

Sir, — No lists of the militia captured at Queenston have come to my hands, or I presume been taken. The printed statute which you mention as being enclosed to n>e formerly I did not receive.

(Tompkins Papers, Vol. VIII., p. 202, NYS Library.)

Letter from Lieut. Patrick McDonogh to his Parents.
Williamsville, December 12th, 1812.

Dear Parents: I received yours of the rot10th this moment and one dated the 17th of last month on the 10th inst. It surprised me to find that Smith, who sutlered for us at Mantua, should' furnish you with my bill, as I expected he was coming after us as he promised, & which indeed prevented me from paying him until I could better spare the money. He has, however found means of adding to the amount as it was originally but 8 dollars, and in saying he lost by the officers of our detachment he said what was infamously false as all but myself settled with him.

He might have come on after us to Albany: but we left him in Philad. I give you my word that I have not recd. a cent of pay since I left you nor do I want money more than to settle what little I owe which I will be enabled to do by the time this reaches you, as we expect the Paymaster the latter end of this month or the beginning of next, when I will forward Mr. Smith's account with the money for the boots and about $15 more that I owe. I assure you it was not for want of principle that I did not pay them before.

We were led to believe we would get money in Albany but were disappointed. My clothes have held out and we can't starve in the army. I commenced drawing my rations the 16th of this month, Keyler cooks for me, and as good living as bread & beef can give, I have.

We are now encamped in the woods, building Huts which we expect to get into by the middle of next month it is rather late in the season to be in tents.

We have a very handsome situation on Elliott's [Elliott's] Creek. The place is called after its owner a Col. Williams of New York I hear he contemplated building his house next spring on the very ground on which we are building and desired that not a piece of timber should be cut down as he wished it entirely shaded, but I can promise him that by that time there will not be a sapling standing within a mile of it. We marched from Black Rock to this place on the 11th inst.

You have heard ere this of the duel that took place on the 12th between Generals Smith & Porter and the latter's statement of facts. Genl. S. has left here on furlough; Col. Porter of the Light Artillery commands in his absence, and as we are 11 miles from the enemy everything is quiet,

Your affectionate son

PATRICK MCDONOGH

I need not mention that it was not I that commanded the

gun-boat; no such good luck for us in this quarter-our hands are tied. Direct to Buffalo, it is the only post town near here-11 miles distant.

(Publications of the Buffalo Historical Society, Volume V p.p. 78-80)

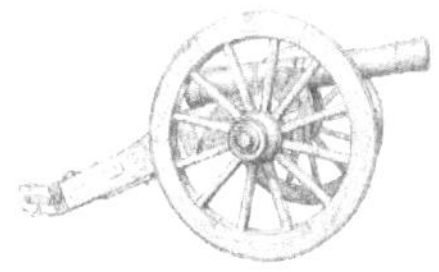

General Alexander Smyth to Governor Tompkins.
Williamsville, 16th Dec. 1812.

Sir, — I understand that Mills, the person you sent to instruct cavalry, has gone over to the British at Schlosser. He took in company the keeper of a public house there and a dragoon. The conduct of this man at Black Rock from the 28th Nov. to the 1st Dec, was suspicious. Nothing but my great respect for Your Excellency could have induced me to give an order recognizing him as an instructor, a man in whose looks and demeanor the villain was so plainly visible. I now mention this as affording a ground for being cautious in sending strangers to situations where they have the most complete opportunities to see our force and learn our intentions.

(Tompkins Papers, Vol. VIII., pp. 302-3, New York State Library.)

From Poulson's American Daily Advertiser of Philadelphia –
January 1813, p. 341:

Captain McKeon of the U.S. Artillery arrived in New York on Tuesday last from the lines. He informs us that he left the army in comfortable winter quarters at Eleven Mile River (Eleven Mile Creek).

Baptist Irvine to General Porter.
Buffalo, January 6th, 1813.

So petty is the force we have in Buffalo that to anything like an army, it is only a picket guard. Two hundred men might destroy everything here if they surprised us. Many, if not all, agree that it was wrong to place the regulars so far off as Eleven Mile Creek where they could afford us no assistance in case of attack.

(From MSS. of Hon. P. A. Porter.)

General Boyd to General P. B. Porter.
Headquarters, Fort George, July 30, 1813.

Sir — As soon as the health of Captain Saunders will admit you will please to remove him to Buffalo or Williamsville. The latter place being more remote will be the most eligible provided it is equally commodious.

(From MSS. of Hon. P. A. Porter.)

Colonel Winfield Scott to General Wilkinson.
Fort George, Monday, 7 o'clock p.m., October 11, 1813.

(Excerpt) Doctor Mann and Captain Camp have concluded to move the general hospital to the Eleven Mile Creek near Buffalo, the barracks at which place will be sufficient for the reception of the whole of the sick with some trifling repairs.

(American State Papers, Military Affairs, Vol. I, p.p. 482-3

General Orders

Head Quarters Batavia, Jan. 14, 1814.
Major Riddle,

Sir: — You are directed to march the regular troops under your command towards Buffalo; you will station yourself at or near Major Millers, wherever you can find convenient quarters for your forces. You will keep a patrol from your camp to Buffalo and Black Rock, and will report yourself to Brig. Gen. Hopkins, commanding at Williamsville, whose orders you will respect. In your march you will keep your men embodied and subject to orders. Circumstances require that your march be commenced very early tomorrow morning and conducted with expedition. Should you discover any movements of the enemy causing an alarm,, you will communicate the intelligence to Gen. Hopkins.

Bv order, etc.

G. H., A. D. C.

(Publications of the Buffalo Historical Society, Volume V p.p. 41-2.)

ANTHONY LAMB TO GOV. TOMPKINS.
ALBANY, Jan. 20th, 1814.

Sir: — Agreeably to your excellency's orders, I left this city on the 4th instant, and proceeded with all possible dispatch to the Niagara frontier. On my arrival at Bloomfield, in the county of Ontario, I met Gen. Hall, who was on the point of returning to Batavia, to collect the troops who had been detached under your excellency's order of the 26th November, which amounted to one thousand, which ho had ordered out on authority, subsequently given him.

On my arrival at Batavia, I found that the inhabitants of that place and the country west as far as Buffalo, on the main road, had, on receiving information of the landing of the enemy, fled and left their homes but were generally returning. I proceeded to Buffalo and found that flourishing village totally destroyed. The only buildings remaining in it are the goal, which is built of stone, a small frame-house and an armorer's shop. All the houses east of Buffalo on the Batavia Road, for two miles, excepting log houses, are also destroyed, and almost every building between Buffalo and Niagara, along the river, had, I was Informed, shared the same fate.

The enemy had with him at Black Kock and Buffalo a number of Indians, (the general opinion in that country is about two hundred,) who pursued their accustomed mode of horrid warfare by tomahawking, scalping and otherwise mutilating those who fell into their hands. Among the victims of their savage barbarity was a Mrs. Lovejoy, of Buffalo, who was tomahawked and afterwards burnt in her own house. The conduct of these savages has struck the minds of the people on the Niagara frontier with such horror as to make it absolutely necessary that a more efficient force than ordinary militia of the country be employed for its protection, to prevent its becoming entirely depopulated.

There was, when I left Batavia, between five and six hundred militia at Williamsville and in its vicinity, under the command of Gen. Hop- kins, and about the same number on the Ridge Road, near the arsenal, under the command of Col. Hopkins. It was the intention of Gen. Hal! who was at Batavia, to make up a number at each of these stations to one thousand men. There was also at Batavia about one hundred regulars, under the command of Major Riddle, who had received orders to march to Williamsville.

As the enemy had re-crossed into Canada, leaving no part of his force on our territory, except the garrison of Fort Niagara, I did not think it necessary for me to remain in the country, or to exercise the plenary powers with which you were pleased to invest me, especially as the authority given to Gen. Hall appeared to me to be amply sufficient to enable him to give a temporary protection to that frontier. I am decidedly of opinion, however, that it is absolutely necessary that a force of a more permanent, substantial nature should be provided, with as little delay as possible.

Sir, I have the honor to be,

With the greatest respect,

Your most obedient servant, ANTHONY LAMB.

His Excellency Gov. Tompkins.

General Orders

Batavia, Jan. 21st, 1814. Major Gen'l Mead

Sir : — This will accompany my letter in answer to yours of the nth inst. I was not informed until last evening that the express who brought your letter was waiting for an answer. I was at that time on the northern part of the frontier and your express came no further than Williamsville. I had ordered an express to go through to Erie to start this morning but shall send this by your express.

I have nothing new to inform you of respecting the movements of the enemy nor can I ascertain to my satis- faction whether their main force is gone on any secret expedition or not. The following is the latest and I believe the most correct account to be obtained fa break in the MS.] except the forces said to be above Chippewa com'd'g the navy at Erie stating their apprehensions of an attack at the latter place as soon as the ice becomes sufficiently strong to pass over which generally happens by the loth of Feby. They have requested me to co-operate with them by station- ing my force at Chatauque. In answer I have been obliged to state that my force on this station will be small and hardly sufficient to guard the frontier and quiet the fears of the Inhabitants. My force may be calculated as follows : First detachment of 1,000 men at 600; second detachment of 1900 say 1000, possibly 1200, making at most 1800 men; this force will be stationed at Williams Ville and near Lewistown, and nearly equally divided. There are about 150 regulars (such as they are) on this frontier without officers except a Lieut. Riddle, who I am informed by himself is ordered to superintend the recruiting service at this place. Major Malcomb arrived this evening with orders from Genl. Wilkinson for those troops to join their several regiments at French Mills,* etc. The second detachment of militia is coming in daily and as fast as they can be organized march on to W. Ville and the cantonment near Lewistown. There being neither axes nor camp kettles in the Q. Master's department I have been obliged to order a partial supply to be purchased by the A. D. Q. M. Gen. at this^ place until a supply shall arrive from Albany which I hope may be soon or the troops will suffer exceedingly.

(This letter is printed as it is in the manuscript. It Has no signature and is obviously incomplete. There is also, it would seem, a considerable loss at the point where a break is indicated, the part before it being on one sheet, that which follows on another. It is even possible that the two portions are parts of different letters.)

(Publications of the Buffalo Historical Society, Volume V p.p. 45-6.)

Head Quarters, Williamsville, Feby. 17, 1814. To His Excellency the Governor

Sir : — I feel it my duty to communicate for the information of your Excellency some circumstances necessary to be known, and which it is thought will require your early attention, and probably the cooperating aid of the Legislature.

In the first moment of alarm on this frontier; for the purpose of defense, and to supply present deficiencies, arms and ammunition were provided by any means, wher-ever they were to be had, of individuals or otherwise. The ammunition has in part been paid for on a requisition, by the O. Masters department of the U. States, but the arms purchased have not been, nor is there any probability that they will be paid for by the U. States. Those arms have been appropriated to the use of the State, and have been turned into the public arsenals, the individuals of whom they were purchased, remain to be paid, and are men who cannot with- out inconvenience remain unremunerated. These purchases were chiefly made on the authority of Lt. Colo. Davis, and although the proceeding was not strictly warranted by law or usage, yet it is presumed your excellency will perceive in the correctness of the motive a fair claim on the State authority for an assumption of such contracts. The sum requisite to cover such purchases will probably amount to 10 "or 1200 Dollars.

Your Excellency will recollect that immediately after my assuming the command on this frontier a representation was made of the destitute situation of our troops as to arms and military stores — an event naturally resulting from the loss of the great depot of the muni-tions of war, Fort Niagara: as well as the other contingent losses of war. This representation received prompt attention, and a supply it is understood was ordered on to the frontier. It would have relieved me from much embarrassment, had there have ac- companied the arms, etc., a bill of them. But learning that they had arrived at Canandaigua, and were placed under the care of Captain Ridge way a U. S. officer, an order was given to remove them to Batavia. The arms have chiefly been delivered out, it is matter of regret that there should have been so great deficiency of cartouch boxes — a very considerable number of the arms were un-supplied in this particular. The importance will readily be perceived of sup- plying this deficiency. Of ammunition and flints there is a sufficiency for the present.

The camp kettles and pans so much desired have been received and distributed by the O. M. of Brigade, under the supposition that it was the property of the State.

I flatter myself that your Excellency will lose no time in ordering detachments of troops from other divisions into service to relieve those at present here. On this so momentous a subject it is earnestly hoped that there may be no delay, the w-hole number now on this frontier do not ex- ceed 1800 men, and the terms of the militia of the 1st detachment called for on the requisition of the War Department wall expire as soon as new troops can be got out. Those of the 2d detachment having been ordered out under the State authority it cannot be thought reasonable should be much longer detained. The harassing duties of the 7th division, call loudly for consideration, and it is hoped that those citizens more happily situated will commiserate and relieve them From recent information the enemy are undoubtedly in considerable force near this frontier, and adequate -security cannot be afforded without considerable addition to the numbers now in service.

There is another subject I am constrained to press upon your Excellencies attention ; the troops which have been and are now in service under the State authority have only the promise of pay expressed on the Statute book,' but it is not known that any appropriations are made by law for their pay. nor is it known that the District Pay-master of the armies of the U. States will feel authorized to pay such troops without express orders from the War Department. Some uneasiness has already been manifested on this subject, and it would afford me much satisfaction to be in- formed by your Excellency as to what answer I can give to those who may inquire of me respecting their pay and the means which may be provided. A satisfactory assurance on this subject may be of important service.

I have the honor to be your Excellencies most obedient and humble servt.

A. HALL.

(Publications of the Buffalo Historical Society, Volume V p.p. 52-4)

General Jacob Brown to General Peter B. Porter.
(Confidential)
Williamsville, April 17th, 1814.

Dear General, —You have seen General Hall's General Order. Such a production would break an officer in the regular service. It is our duty and I suppose the duty of every officer to endeavor to give character and strength to the Government and country. Any attempt on the part of a person that the Government has authorized to wear a sword to destroy all confidence in that Government appears to me to be a high military offense.

General Hall stated to me that it was impossible to retain the militia in service. He had discharged the whole of those that were at this place before my arrival, notwithstanding I had particularly desired him to retain the last draft, if those whose time of service had expired could not be prevailed on to stay.

You will consider whatever I may say to you on military subjects as confidential, as I desire to conceal nothing from you.

No orders have yet reached me from the Secretary of War. But I know that he is well placed with the mistake, and I do not doubt but the necessary arraignments will be made for crossing (a little below Long Point, (with the aid of our flotilla on Lake Erie,) and by turning the enemy's position attacking them in rear. Of this, more when I see vou.

(From MSS. of Hon. P. A. Porter.)

General Peter B. Porter to General Brown.
Canandaigua, April 21, 1814.

Sir, the bearers are three young warriors of the Seneca nation on their return from Canandaigua, where they were sent to obtain some explanations from Mr. Parrish and myself relative to the late Talk of the Secretary of War and are to refer our answer to a council of chiefs now sitting at Buffalo. Before the council is dissolved they are to make a final decision on the propositions of the Secretary, which I presume will be to accede to them, and we have reassured them to put themselves in a state of complete preparation so that they may assemble at a day's notice.

The Onondagas and Oneidas have lately held a general council and agreed that they will all turn out and assist their white brethren to drive the British from our land, The result of the council was transmitted to this place by a delegation of warriors, and will be conveyed by the elder's hereof to the Senecas at Buffalo.

My prospects as to volunteers are yet promising, but I cannot calculate with any certainty on the number that will be raised.

My opinion of General Hall's valedictory order agrees with ^ours. I had no doubt that he possessed the dispositions manifested m this order, but I had hardly expected to see them exhibited in this way.

I entirely concur in the policy of the measures intimated in the latter part of your letter, and, as you wish, shall consider it strictly confidential.

P. S. —It is customary to give provisions to the Indians when assembled on public business, especially if convened at the request of the Government. As Mr. Granger, (the agent,) is now absent, I presume you will order bread and meat to be issued to the chiefs now at Buffalo. The bearers will probably be out of provisions by the time they arrive at Williamsville.

(From MSS. of Hon. P. A. Porter.)

From the Narrative of Alexander McMullen, A Private Soldier in Colonel Fenton's Regiment of Pennsylvania Volunteers.

Wednesday July 6, 1814—Some of the men of Col. Fenton's detachment who had remained and had not been furnished with rations, but at their own expense, until now, began to cross over and follow the detachment, and others who remained were ordered to attach themselves to Capt. Piper's company, the Provost Guard at Buffalo. Some New York Volunteers arrived in Buffalo this day, and drew muskets; their number was 320, generally boys from 12 to 16 years of age, in the afternoon the New Yorkers consisting of the above mentioned 120 Light Horse and 320 footmen, crossed over at Black Rock and marched to join Gen. Brown. At 3 o'clock the Carlisle Infantry marched with prisoners for Greenbush, the company consisted of about 70 men fit for duty, 10 other volunteers who had not crossed over were expected of to assist in guarding the prisoners who were above 130 in number. The day warm and the roads very dusty, we arrived at a village called Williamsville, on Eleven-mile creek at sundown —11 miles.

Thursday July 7, 1814—There are Barracks or rather huts of wood in Williamsville in which we lay the preceding night, there are two heavy guns placed in the center of the village for its defense. A company of Porter's Volunteers, about 60 in number, arrived here this morning on their way to Buffalo. We marched at 6 A.M. The day was very warm, but the road is level and the country handsome, as it is all the way from Buffalo to Batavia at 7 P.M. The prisoners began to grumble about our hard marching, or we should probably have reached Batavia to day—25 miles.

General E. P. Gaines to General Porter.
Headquarters, Fort Erie, U. C,
August 16th, 1814.

Sir, — You are hereby authorized to accept for the U. S. the services of four thousand volunteers, or less, for one, two or three months, unless sooner discharged by proper authority, to rendezvous, be organized, mustered, inspected, and provisioned at or near Williamsville, as soon as practicable ; to have the choice of their own officers according to law, and be subject to the same rules and regulations, receive the same pay, and be provided for in case of being killed or wounded, &c., as United States volunteers, agreeable to law. This corps will not be ordered on duty unusual or improper for this description of force. I shall not undertake to persuade my young countrymen to come out to the tented field. They have heard the eloquence of our cannon for some day's past. This, I am sure, is fully sufficient to excite their military ardor and call them to the field, where American troops if true to themselves will never fail to reap honorable victory over foreign slaves and mercenaries. (From MSS. of Hon. P. A. Porter.)

Buffalo Gazette, Tuesday, August 16, 1814.

Two companies of the 19th Regiment arrived in Buffalo on Thursday evening last from Erie, Pa. The detached militia, under Lieutenant-Colonel Brooks, are stationed at Williamsville.

State of New York— General Order.

The Commander-in-Chief, under and pursuant to the Act of Congress passed 28th February, 1795, and the Acts supplementary and additional thereto, directs the following detachments of infantry to be made from the division commanded by Major-General Hall, exclusive of the requisition made by Major-General Brown for one thousand men, which Major-General Hall is also required to complete.

The detachment required by this order will be made without any delay, and are directed to rendezvous at Williamsville, in the County of Niagara, on the 1st day of September next, where they will be organized :

From the brigade of General Tillotson, Cayuga county, 1,000 men.

General Graham, Seneca county, 350.

General Burnet, east part of Ontario, 560.

General Wadsworth, west part of Ontario, 560.

General McClure, Steuben county 350.

General Davis, Genesee county, 180.

And all the companies of grenadiers, riflemen, and light infantry within the counties of Cayuga and Seneca, together with the artillery companies in Auburn and Geneseo, are also directed to assemble at the same place of rendezvous on the first day of September next. Independent companies, companies of exempts, and other associations who tender their services are to be accepted and organized by General Peter B. Porter, who will command the whole detachment.

Every soldier must furnish himself with a musket or rifle, knapsack, canteen, cartridge box, three flints, a watch coat, and clothing for three months. Those who are unable to equip themselves with muskets and rifles or cartridge boxes will be supplied from the public deposit, but it is required by the Commander-in- Chief that all who can supply themselves should do so. The militia thus detached will be organized into companies of 100, including one captain, two lieutenants, two ensigns, five sergeants, six corporals, one drummer, and one fifer to each company. The companies will be formed into regiments of 1,000 men as near as may be, including officers, with two lieutenant-colonels and two majors as field officers. Brigadier-General Peter B. Porter will have the immediate command of the whole detachment, subject to the orders and directions of the Commander-in-Chief. Lieutenant-Colonels Hugh W. Dobbin, Peter Allen, Caleb Hopkins, George W. Flemming, and James McBurney, and Majors Thomas Lee, Jr., Daniel Cruger, Eranthus Everts, and Eben. Willson of Genesee are assigned to the respective regiments in this detachment; the remainder of the field and staff officers will be assigned by Brigadier-General Peter B, Porter. The captains and subalterns will be assigned by the commandants of the respective brigades. The independent uniform companies will be commanded by their own officers or by so many of their own officers as will be in proportion to the number of men, and if, on their arrival at the place of rendezvous, the officers of any corps be more than in proportion to the number of men, such supernumerary officers as the commandant of the whole detachment shall direct will be discharged. The troops will receive the same accommodation and pay as are provided for the army. The Commander-in-Chief feels assured that the officers, non- commissioned officers and privates will evince the greatest alacrity in complying with this order. The present situation of our country loudly demands vigorous exertions. The Niagara frontier presents a scene which must necessarily arouse the feelings of Americans and urge them on to the relief of their countrymen. On one side of the river an important American post in possession of the enemy — on the other, Fort Erie occupied by the American army. The latter pressed by a superior force without the power of retreating. Too great a delay may enable the enemy to reinforce in such numbers as to become irresistible to the troops now across the Niagara, and that gallant little army, the hardy remnant of several well fought battles, in which they have as often defeated as they have engaged their foe, will fall a sacrifice f>o the want of patriotism and philanthropy, to the inertness and the apathy of their fellow citizens. Should this army be destroyed the whole of the western frontier of this State will be exposed to the ravages, the devastation, the burnings, and the murders of a merciless foe.

The Commander-in-Chief has, however, the utmost confidence that not a man will be called on at the present time whose arm will not be raised in defense of his country, and who will not feel the responsibility of his conduct and bring a corresponding zeal to the place where his duty and the safety of his fellow-citizens require his presence.

By order of the Commander-in-Chief.

John B. Yates,

Lieut.-Col. and Aid-de-Camp.

(From Buffalo Gazette, August 30, I814)

General George McClure to General P. B. Porter.
Bath, 26th August 1814.

Dear Sir, My Brigade Major, John Wilson, is anxious to be employed. He in a vigilant and active officer. Should you think proper to assign him a command you will please inform him or me by express mail.

The detachment from my brigade is ordered to assemble at this place on the 30th inst. and will proceed with all possible speed to Williamsville via Canandaigua.

I fear there will be but a small proportion of them appear on the day of rendezvous, and unless some effectual mode of punishing delinquents is speedily adopted future calls on militia will be useless.

Such is the language of those who have and are still willing to do their duty. I sincerely congratulate you on your good success so far. I trust the campaign will tenninate much to your honor and the glory of our American arms. Should Chancey continue to ride triumphant on the lake we would have but little fear on our frontier.

(From MSS. of Hon. P. A. Porter.)

General Brown to General P. B. Porter.
Headquarters, Fort Erie, ' September 8th, 1814.

Dear Sir, — I have ordered every tent from Buffalo, and that the sick and wounded be put into the houses. In times like these we must look to nothing but the great interests of our country, and there- fore I shall this day also order that the sick and wounded at Williamsville he also put into the nearest houses that can be found, and the tents forwarded to this place. It is painful to me to be compelled to resort to this measure, but I find it my duty, as I have no other means of covering the men under your command.

You will order over the best regiment you have as soon as possible; the tents from Buffalo will cover them. Let me know when we may expect your whole force. My opinion is that we shall not gain by delay as it may relate to your command. General Izard is said to be in motion. I shall be better informed upon this subject in the course of a day or two. Send over a regiment to-day if practicable; if not today, as soon as possible — say to-night or to-morrow.

(From MSS. of Hon. P. A. Porter.)

General Orders.
Buffalo, Sept. 11th, 1814.

On Gen'l. Porter's leaving Buffalo this morning, Maj. Cruger will take command of the post.

Major Cruger will consolidate and organize the remnants of companies left on this side, assign to them proper officers, arm such as are destitute, and put them in a state of preparation for active operations as speedily as possible.

The sick of the several corps of volunteers and militia will lie placed in some convenient houses, and surgeons assigned to take charge of them.

Major Cruger will send a company of 60 men, with proper officers, to Williamsville, to report to the commanding officer and remain there as a guard. The British prisoners and deserters in the provost will be sent to Williamsville undercharge of this guard. The provost marshal will furnish a list of their names.

Maj. Cruger will station piquet's at B. Rock and Miller's and at such other places as he may judge most proper for the security of the post and to take up deserters. In short, Gen'l. Porter relies on the Major to place the distracted materials now left under his com- mand in the best and most effective state and have them ready to march. He will send across to Erie from time to time such as may volunteer to go there. Major Cady will organize the militia of Gen'l. Davis's brigade, consulting with Major Cruger.

(From MSS. of Hon. P. A. Porter.)

General Izard to General Porter.
Headquarters. Buffalo, November 1st, 1814.

Major-General Porter, New York Militia:

Dear Sir, — If you could impress the principles on which the papers enclosed in your note of the 28th October are predicated on the people of our country at large, you would do more towards bringing the war to a speedy and honorable termination than all the acts of Congress and votes of town meetings from one end of the United States to the other. As to Captain Watson's company. I had heard of its flight before I crossed the straight, and have substituted a company of regulars, who are now at work there. If some of the intelligent men among them would volunteer to assist us it would [torn] much credit and us much service, [torn] can be done; have the goodness to direct them to report to Doctor Bull, hospital surgeon at Williamsville.

(From MSS. of Hon. Peter A. Porter.)

Major-General Izard to the Secretary of War.
Buffalo, November 8th. 1814.

Sir, On the 3rd instant I had notice that the British fleet was again at the mouth of the Niagara, and landing troops and stores. My informant stated several pieces of heavy ordnance and fifteen hundred men to have passed that morning from Fort George to Queenston Heights. The enemy, however, advanced no force beyond Frenchman's Creek. All appeared quiet in the vicinity of this position. On the morning of the 5th, everything having been removed from Fort Erie and the preparations for its destruction having been completed, the small remaining garrison was embarked, and soon after the bastions were blown up; it is thoroughly demolished. The enemy was not aware of what had been done for more than twenty-four hours afterwards. The whole of my force is now in the vicinity of this village engaged in constructing huts. The scarcity of forage has induced me to send the dragoons to the Genesee country, where they will be kept better and at one fourth of the expense. I have dispatched Lieutenant Colonel Ball and an officer of the Quartermaster General's department to select the ground for their winter quarters, and in the meantime, they have moved to Williamsville, where there is provender sufficient for a few days. The construction of the hospital at the latter place has been delayed by the jealousy and quarrels between the surgeons and the Quartermaster's department of General Brown's division. It is far from ready for the reception of the sick and wounded, who remain principally under tents. By the advice of Dr. Bull, senior surgeon, I have directed such as could bear the transportation to be conveyed to Greenbush in wagons; among these will be the wounded British prisoners. But the remaining number is distressingly great; there are yet nearly two thousand men on the surgeon's list. No information has reached me from the westward since my last to you, of the 2d instant. If the flotilla were here, I would immediately detach a thousand or twelve hundred men to Detroit by land it is now impracticable. The dispersion of troops assembled with so much trouble and difficulty would be an unpleasant measure, but I know no other means of protecting Detroit from an attack, which seems to me not an improbable event. The enemy will, I believe, notwithstanding their now very superior numbers, be disposed to let this part of the frontier alone. With such a fair opportunity as was repeatedly offered them on the Canada side of the strait, I doubt their inclination to resume the offensive, especially in this ruined country.

There is, however, a greater probability of their sending a force to the westward. On the subject of the preparations by General McArthur and Governor Cass, I am entirely destitute of information. The troops under my immediate command, being now engaged in constructing their huts and the final disposition being made of the various detachments and garrisons, there will remain in a short time nothing on this frontier to require my presence. The attention of the government will probably be immediately directed to the operations of the next campaign. If it is the intention of the President to honor me with the command of the 9th District during the next season, it will be all important that I should employ the winter months in making such arrangement as will enable us to take the field and seize some strong positions early in the spring. Much is to be done in every department of the army. System must be enforced in the various branches of ordnance, clothing, quartermaster's, surgeon's, and pay departments. Incalculable expense is produced by the present disorganized state of things. The staff of the army is on a very unpleasant footing. No regular course of discipline and instruction is followed by the troops.

Much may be done in the next five or six months, but it must be quickly commenced to be of permanent advantage. I could, at Philadelphia, communicate more rapidly with the Department of War and there superintend the various supplies required for the frontier. At the same time, I should have an opportunity of paying some attention to my health, which has suffered considerably by the two last campaigns, and the restoration of which I can scarcely hope for if I remain the winter in this inclement climate.

(From Izard's Official Correspondence)

Report of Hospital Surgeon Lovell of the State of Diseases among The Troops on the Niagara Frontier During the Campaign of 1814.

The troops engaged in this brilliant campaign on the Niagara began to collect there about the beginning of April, under the command of General Scott. They were encamped on an eminence north of Buffalo

village, having a thick wood in front which extended to the bank of the river, the ground being in part swampy and wet. On the left of the encampment was a large marsh extending from the high ground to the margin of the lake. The winds from the lake at this season were remarkably cold and chilling; resembling in sensation exactly the east winds which prevail on the Atlantic during the spring, and have an astonishing effect upon vegetation, the trees around the encampment having the appearance of winter, while

those five or six miles from the lake shore were covered with verdure. Notwithstanding this, the troops were remarkably healthy, only one or two deaths occurring before they crossed the Niagara on the 3rd of July even the demon diarrhea appeared to have been exorcised by the mystical power of strict discipline and rigid police.

In June a number of new recruits joined the army, and several were collected from the various hospitals, the latter principally composed of the miserable refuse of society who never had energy to demonstrate that they lived, and scarcely enough to prove that they existed. With these last detachments arrived our old acquaintances, which, however, were easily checked, and much soldier returned than in any former campaign. This was undoubtedly to be attributed to the improvement in police.

During June the weather became very warm, and a thick fog arose from the marsh and woods at sunset and remained for some time after sunrise During this month intermittent, acute rheumatism and typhus fever were the prevailing complaints. The intermitfcents were very irregular and obstinate.

Rheumatism during the whole war generally put on a remitting form. This was particularly obvious whenever intermittent fever prevailed, and more especially this season.

Many of the cases of typhus, about the end of May, were remarkably severe.

On the first of August a general hospital was established at Williamsville, eleven miles east from Buffalo. The number of sick during the remainder oi the season at this place varied from 3 to 400; the number of wounded being somewhat greater.

The troops suffered much during the siege of Fort Erie, and soon after it was raised the rainy season commenced. Dysentery and diarrhea were the principal diseases.

About the end of September, a large detachment of militia crossed the Niagara under General P. B. Porter. Diarrhea, typhus and idiopathic dysentery very soon made their appearance among them; the two latter were extremely severe.

(From Mann's Medical History of the War)

References to The Williamsville Hospital
Medical Sketches of the Campaigns of 1812, 13, 14
By James Mann M.D.A.A.S Printed 1816

From page 94.

These changes determined me after spending some days to select a secure position and prepare quarters for the approaching winter to break up the encampment at Lewistown We had at this time about three hundred and fifty in the hospital including the sick convalescents and invalids The last were subjects for discharge in consequence of wounds and other disabilities After selecting one hundred who had so far recovered from wounds and bodily infirmities as to be able to perform a journey of three hundred miles to Greenbush the remainder were ordered to Eleven Mile Creek called Williamsville These arrangements were communicated to Colonel now General Scott who had been left in command at Fort George and met with his full approbation To the assistant quarter master General of the post he gave his order to furnish the necessary means of transportation The invalids destined for Greenbush were committed to the charge of Lieutenant ARCHER regimental pay master and Doctor Hugo The remainder of the sick and convalescents two hundred and fifty in number some of whom very debilitated were transported to Schlosser a distance of eight miles in wagons thence to Black Rock in boats from the last place in wagons to Williamsville This was a tedious journey of forty miles for sick men many of whom were destitute of clothing suitable for the approaching inclement season.

From page 95.

At Williamsville the barracks which the preceding winter had been occupied by General SMYTHE's division were put into a comfortable state of repair and improved during this winter as hospitals These quarters were very extensive and were by much labor rendered commodious for the number who composed this detachment of sick and convalescents The wards were made warm this was more important as the men were destitute of clothing. The wards were not crowded consequently were less liable to become infectious. Six patients only were and at placed in a room wherefore it required but little labor and attention to preserve them in a state of cleanliness. This part of duty the soldiers were at all times obliged punctually to execute Those regulations which had respect to cleanliness were always scrupulously enforced within the hospitals under my direction.

From page 96.

At Williamsville before three weeks had expired fifty of the convalescents were reported fit for duty and were ordered to Fort Niagara. The barbarous deaths of most of these men with the rest of that ill-fated garrison we have to lament while emotions of just indignation irresistibly obtrude themselves when reflecting upon the manner. Surprised and taken without resistance most of the garrison were deliberately murdered by the bayonet after surrender. Having disposed of the sick in quarters for the winter in conformity to the orders received from General WILKINSON and finding their number daily decreasing my services at Williamsville were believed to be no longer absolutely necessary.

From page 97.

The charge of the hospital at Williamsville was resigned to Doctor WHITRIDGE who for assiduous attention to duty was exceeded by no physician of the army. The services of this gentleman cannot be too highly appreciated. To him not only the nation is indebted, but my grateful acknowledgments are due, and this opportunity is improved more fully to express them. A full share of the high reputation which the hospital at Lewistown supported during the campaign as announced in the flattering reports of the Inspector General to the Commander in Chief of the army is attributed to his fidelity and good judgment.

From page 112.

The last week of November I departed from the Niagara frontier to follow the division of General Wilkinson After leaving Williamsville thirty miles east we arrive at Batavia a new well-built village surrounded with farms extensively opened comprising a soil of the best quality The village contains 40 houses a handsome court house jail and a county arsenal.

From page 162.

On the first of August a general hospital was established at Williamsville eleven miles east from Buffalo The number of sick during the remainder of the season at this place varied from 3 to 400 the number of wounded being somewhat greater.

From page 251.

The movement of 250 sick from Lewistown to Williamsville a distance of 40 miles was painful to the spectator it aggravated the disorders of many and caused the death of some. This was performed the last of October There were but two alternatives either to erect hospitals from the rough timber of the forests in the vicinity of Lewistown or occupy the barracks built at Williamsville (eleven mile Creek) by General Smyth's division The first required too much time to complete at this advanced season of autumn the last was considered the most eligible as the men would have less hardships to endure by a remove to houses already erected than by necessary exposure while new huts were building.

The barracks at Williamsville wanted considerable repairs but in a short period were rendered comfortable During this remove Doctor WHITRIDGE was my only assistant.

**From *The Early Firm of Juba Storrs & Co.*,
Rev. Albert Bigelow:**

The declaration of war at once brought business to the young firm, especially at Williamsville. They made a contract with the government for all the mill products they could furnish for the army and had work enough to do. There were no other mills nearer than Niagara Falls, except, as Lucius Storrs with a somewhat poetic license was wont to say, one that wasn't bigger than a coffee mill. Then, in the winter of 1812-1813, barracks and a hospital were established about a mile from Williamsville up the creek, by cutting down trees and building huts in the woods.

Before these were built, one regiment of troops were cantoned in front of Williams' house: afterwards another behind the house, down the creek. On the bank of the creek was a beautiful grove, and the engineers asked permission to cut down trees, to build barracks with. But this Mr. Caryl refused as unnecessary destruction. They replied, however, that his refusal made no difference--their asking was a mere matter of courtesy—and so they went to work in spite of his refusal and cut down the trees.

Before the barracks were built, a temporary hospital had been established. In the tavern, Gen. Brown and Gen. Ripley and a British officer were sick, and the men were taken care of in the houses of the little village. And the sawmill of Juba Storrs & Co. furnished the lumber of which the boats were built that carried the troops across the river.

Thus, on account of the mills and hospital, Williamsville was an important point in the war, and large rewards were offered to the Indians by the British for the burning of them. But they were afraid of being intercepted and cut off, so never attempted to burn them.

The troops used these quarters until about April 1813, when they joined the battle force attacking York (Toronto) on April 27, 1813. The huts were pretty much abandoned until October 1813, when it was decided to use the cabins as a winter hospital.

U.S. Army Medical Department Office of Medical History

The U.S. Army Medical department describes the attack at York and the casualties as follows:

On the twenty-seventh of April, the attack on Little York took place, and after four days occupation of the town, the army with the wounded and sick was moved to Fort Niagara, where a tent hospital was organized two miles from the river. The ground was wet and low, and many of the wounded died from camp diarrhea and typhus fever, and in June, after the capture of Fort George, a general hospital was established at Lewistown, eight miles up the river, by advice of Surgeon Mann, Medical Director. This hospital consisted of two barns, besides a large number of hospital tents, and was well supplied with everything necessary for the comfort of the sick and wounded, which, by the first of August, had increased to nearly seven hundred. Here the patients improved very rapidly, the position of the hospital being salubrious, the tents policed with great care, and the diet being generous. In the army at Fort George, however, a most lamentable degree of sickness prevailed.

Confusion exists concerning the nature of the medical facilities in the Fort Niagara-Fort George area. Dr. Mann described a tent hospital located two miles east of Fort Niagara where the wounded from the attack on Toronto were cared for beginning on 8 May 1813 and also noted that about two hundred wounded were moved from Fort Niagara to Lewiston in mid-June 1813.

There was also a general hospital at Fort George, which was taken from the British in late May 1813 and held for several months. In June, some of the patients at Fort George were also moved to the higher, healthier facility at Lewiston. By August, however, more than one-third of the men still there were sick, and with half the medical staff too ill to work, only three surgeons and four mates were available to care for six hundred to seven hundred patients. Hospital stores were running short by September when Mann decided to move the general hospital from Fort George to a site near Buffalo and to send all 100 invalids ready for discharge with a surgeon's mate from Fort George to Greenbush.

During the last weeks of October 1813, it was decided to remove the sick and wounded from Lewiston because of the approaching winter and the dangerous situation. Two hundred and fifty men were transported to Fort Schlossler (Niagara Falls) from where they were taken up river by boat to Black Rock, then to Williamsville by wagon. There, the barracks, which had been erected by Smyth's army, was improved and used as a hospital, six patients to each cabin. Dr. James Mann was appointed surgeon in charge, later to be succeeded by Dr. Whitridge. Having settled his patients in Williamsville's wholesome accommodations, Mann set out to join General Wilkinson at his winter camp at Malone, leaving surgeon's mate Joshua Whiteridge, a physician whose "services cannot be too highly appreciated" in charge.

While Washington Burned: The Battle of Fort Erie 1814, Joseph Whitehorne

[21]. LC, Amasa Troubridge Papers, ibid

Before the invasion, General Scott had periodically visited all the various regimental hospitals, warning the physicians that the campaign would bring busy times for them.[21]

[22]. Robert W Bingham "The History of Williamsville" Niagara Frontier Miscellany xxxiv, (Buffalo Historical Society, 1947): pp. 100-101; Carolyn Shrauger et al, Williamsville NY Where the past is Present (Village of Williamsville Historical Society, 1985) p 14.

His predictions were all too accurate. These field hospitals were supported by an existing hospital system. By 1814, general hospitals already were established in Burlington, Vermont, Plattsburgh, Malone and Greenbush, New York. The threat to facilities at Fort Schlosser convinced General Brown in late July to erect a permanent general hospital at Williamsville, about 12 miles east of Buffalo. Williamsville was the headquarters of the 5th Brigade of the New York State Militia. It had been used by the Army ever since General Smyth had established winter quarters there in 1812.

At that time, log barracks were built along the south side of the village's Main Street between Ellicott Creek and the Garrison Road. These barracks were converted into temporary medical facilities in October 1813. Patients from Lewiston were transported by boat to Fort Schlosser, then overland. Those capable of the trip were then sent to convalesce at the Greenbush hospital near Albany. The Buffalo area hospital was commanded by Dr. James Mann and Surgeons Mate Joshua B Whiteridge. [22]

[23]. NA, R G 94 Entry 407, Box 90, Hogan Vouchers.

The site for the new (General) hospital was selected on 29 July (1814) on the recommendation of senior surgeon, Doctor Ezekiah Bull. Thus, 90 acres and the stables of Raphael Cook's farm were leased for the construction of a general hospital. Ironically, the contract specified "no burying place in the premises" Despite this, a well kept series of mass graves along Aero Drive remain as the last vestiges of this hospital. These further provide evidence that having a burying ground adjacent to a medical facility was an accepted practice. The Williamsville Cemetery contains the remains of both US and British personnel. The British remains are in their own grave on one edge of the plot. The segregation of enemy and friendly dead has been a tacit, if not official, practice on both sides. The Williamsville facility was designated a general hospital in August, and Surgeon's Mate Joseph Lovell was assigned to its command. He was succeeded in 1815 by Doctor William Thomas.[23]

**[24]. NA, R G 94 Entry 125, Box 134 Vouchers;
Percy M Ashburn.**

As soon as General Brown decided to concentrate patients at Williamsville, a vigorous building project was begun. At first, a large tent city was erected as indicated by the use of

3,000 board feet of timber for flooring 100 hospital tents and 12 loads of hay for bed ticking. Each tent could hold 16 to 18 men. In early September, General Brown, once more in command, embarked on his extensive effort to shelter the troops at Fort Erie.

Additional amounts of timber were sent across the Niagara to make tent flooring. Likewise, canvas was delivered for use by the besieged troops. Brown directed that the wounded be moved into permanent buildings so that every tent possible would be available for troops in the field. Captain John Larkin was named supervisory quartermaster for the construction of the Williamsville Hospital. He brought in skilled workmen from as far away as Rochester and Utica. Huge quantities of locally produced brick were purchased in Buffalo and hauled to Williamsville to help construct the hospital barracks. Subsequent purchases of glass and shingles to complete the buildings substantiate the permanent nature of the structures. Patients, among them Generals Scott and Riall, began being moved to the Williamsville site on 30 July.[24]

[25]. Percy M Ashburn, "History of the Medical department of the United States Army" (1929) pp. 33-34. Severance, ibid, p 247. "American Army Hospitals of the Revolution and the War of 1812" Bulletin of John Hopkins Hospital 46, pg 32

An additional general hospital was opened in Buffalo in July to accommodate the surge of wounded from the Battle of Chippewa. This was located at Sandy Town (The Front) 400 yards from Buffalo Creek. The wounded were brought to the site by boat then carried by litter the last few hundred yards. By 1 August, it held nearly 1,100 patients. The British raid on 3 August a few miles north at Scajaquada Creek demonstrated the hospital's vulnerability. Consequently, as many men as possible were removed to the growing Williamsville facility. Doctor Bull and Surgeons Mates Thomas and Lovell supervised the Williamsville facility. Surgeons Mate William E Horner remained at Buffalo caring for the small number of patients who could not be moved. Thereafter, the Buffalo Hospital served as the clearing center for casualties from Fort Erie sending patients to Williamsville as quickly as it could. Those who died there were buried in an adjacent graveyard, identified only years later when human remains were discovered.[25]

[26]. See annex K NA R G 94 Entry 125, Box 134; RG 94, Entry 225, Box 651, Medical Department, Francis LeBaron letter 14 November 1814 to Callender Irvin

Several smaller regimental hospitals in the Buffalo and Black Rock areas had come in with their units. Despite their regimental affiliation, they supported all troops sent to the area. Buildings were leased to house these facilities. In addition, rooms in private houses were rented for the use of convalescent officers. Beginning in October, the sick and wounded at

Williamsville who could travel were sent in a series of convoys to the hospital at Greenbush. Beginning on 8 November, everyone possible was moved to Williamsville. Most officers and about 80 critical patients remained in private houses around Buffalo. The unprecedented casualty rate experienced throughout the campaign combined with that of Plattsburgh and elsewhere to place heavy demands on the medical supply system.

Apothecary General Francis LeBaron at Albany advised his superiors in Philadelphia that the Williamsville Hospital "devoured" reserve stocks "like cormorants" so that by November nothing was left on hand.[26]

Winfield Scott The Soldier and the Man
by Charles Winslow Elliott

In the crowded base hospital at Williamsville*, where hundreds of the mangled survivors of Chippewa and Lundy's Lane lay suffering in the stifling tents, there was no lack of exciting news to keep temperatures up and retard recoveries. Glorious and shameful by turns, the war tidings which reached the patients daily were little calculated to hasten the convalescence of such ardent spirits as Scott, who suffered mental agonies at his enforced absence from the field in addition to the terrible physical torments of his injuries. It speaks volumes for that iron constitution that he survived at all.

*This General Hospital was established on August 1st. It accommodated from 300 to 400 sick and a somewhat larger number of wounded. Joseph Lovell, M.D., surgeon, 9th Infantry, was the hospital surgeon in charge.

War of 1812 Death Register, Jack Bilow:
Recorded Deaths at the Hospital at Williamsville.

Deaths 1812 - Oct–1, Nov–4, Dec -8 (total 11)

Deaths 1813–Jan–9, Feb–3, Mar–4, Apr–2, May–0, Jun–1, July–0, Aug–0, Sep–1, Oct–0, Nov–2, Dec–4 (total 25)

Deaths at Williamsville 1814 - Jan–11, Feb–4, Mar–11, Apr–5, May–8, Jun–1, July–2, (total 42)

During the first half of 1814 there were forty-two deaths listed at Williamsville. But after the battles of Chippewa, Lundy's Lane and Fort Erie, the number of wounded and sick became immense.

Deaths at Williamsville 1814 Aug–40, Sep–75, Oct–69, Nov–36, Dec–44 (total 264)

Deaths at Williamsville 1815 Jan–49, Feb–28, Mar–7, Apr–3, May–2, Jun–4, Jul–1, Oct–1 (total–95)

Other deaths no dates 27, (total 464)

Bingham, Robert W.
Names of the United States Soldiers Buried in the Old Military Cemetery in Williamsville, New York Niagara Frontier Miscellany xxxiv,
Buffalo Historical Society, 1947, p. 106.

Names of Soldiers from the Williamsville General Hospital Logs and other British Prisoners only known to God.

(See Chapter 6 for a detailed list of the dead)

6

THE GARRISON BURIAL GROUND

A history of the War of 1812 cemetery on Aero Drive in the Town of Cheektowaga, NY,

According to *Snake Hill: An Investigation of a Military Cemetery from the War of 1812*, the site selected for the new general hospital was chosen on July 29, 1814, based on the recommendation of Senior Surgeon Dr. Ezekiel W. Bull. A total of 90 acres, including the stables of Raphael Cook's farm, were leased for the hospital's construction. Interestingly, the lease contract explicitly prohibited the establishment of a burial site on the premises. The original wording reads: "No burying place be suffered on the premises."

It is likely this stipulation stemmed from the fact that Raphael Cook himself leased the land from the Frick family and was therefore unable to authorize permanent use of the land such as a burial ground beyond the term of his lease. A similar condition may have applied earlier when Mr. Frick hosted the first hospital on his own property during the summer of 1813, where Artillery Huts served as makeshift medical facilities.

The earliest documented acquisition of the land encompassing the military cemetery in Lot 7, Township 11, Range 7, occurred on October 4, 1808. This tract including Lots 6 and 7, totaling approximately 202 acres was purchased by Frank Ludlow from the Holland Land Company.

Records indicate that Ludlow maintained ownership until January 11, 1816, when he forfeited the land due to nonpayment. There is no evidence suggesting Ludlow ever occupied or developed the land, which may explain why the U.S. Army selected it because it was vacant and unimproved.

On the same date, January 11, 1816, Wilhem Willink and his associates from the Holland Land Company sold the parcel including the site of the cemetery to Benjamin Barton. This transaction is officially recorded in the Clerk's Office (Liber 3, Page 133) and represents the first documented sale of the land containing the cemetery, although the deed makes no mention of a burial ground. See Figure 1 for the original deed.

Subsequent transactions include:

October 17, 1821: Barton sold the property to Thomas Coe (Liber 7, Page 207). See Figures 2 & 3.

July 4, 1851: William Coe transferred the land to John Haskell (Liber 172, Page 459). See Figure 4.

August 6, 1851: John Haskell sold Parcel A (the cemetery property) to John Bliss and John R. Evans (Liber 123, Page 19), with the stipulation that "the ground should not be disturbed or encroached upon." This marked the final private ownership of the cemetery. See Figure 5.

July 12, 1898: The estates of Bliss and Evans deeded Parcel A to the Buffalo Historical Society (Liber 833, Page 283), where it remained under the stewardship of the Buffalo History Museum until its transfer to the Town of Cheektowaga. See Figures 6, 7 & 8.

March, 1985 a quit-claim deed is signed conveying the cemetery (parcel A) from the Buffalo and Erie County Historical Society to the Town of Cheektowaga. (Liber 9444 page 450)

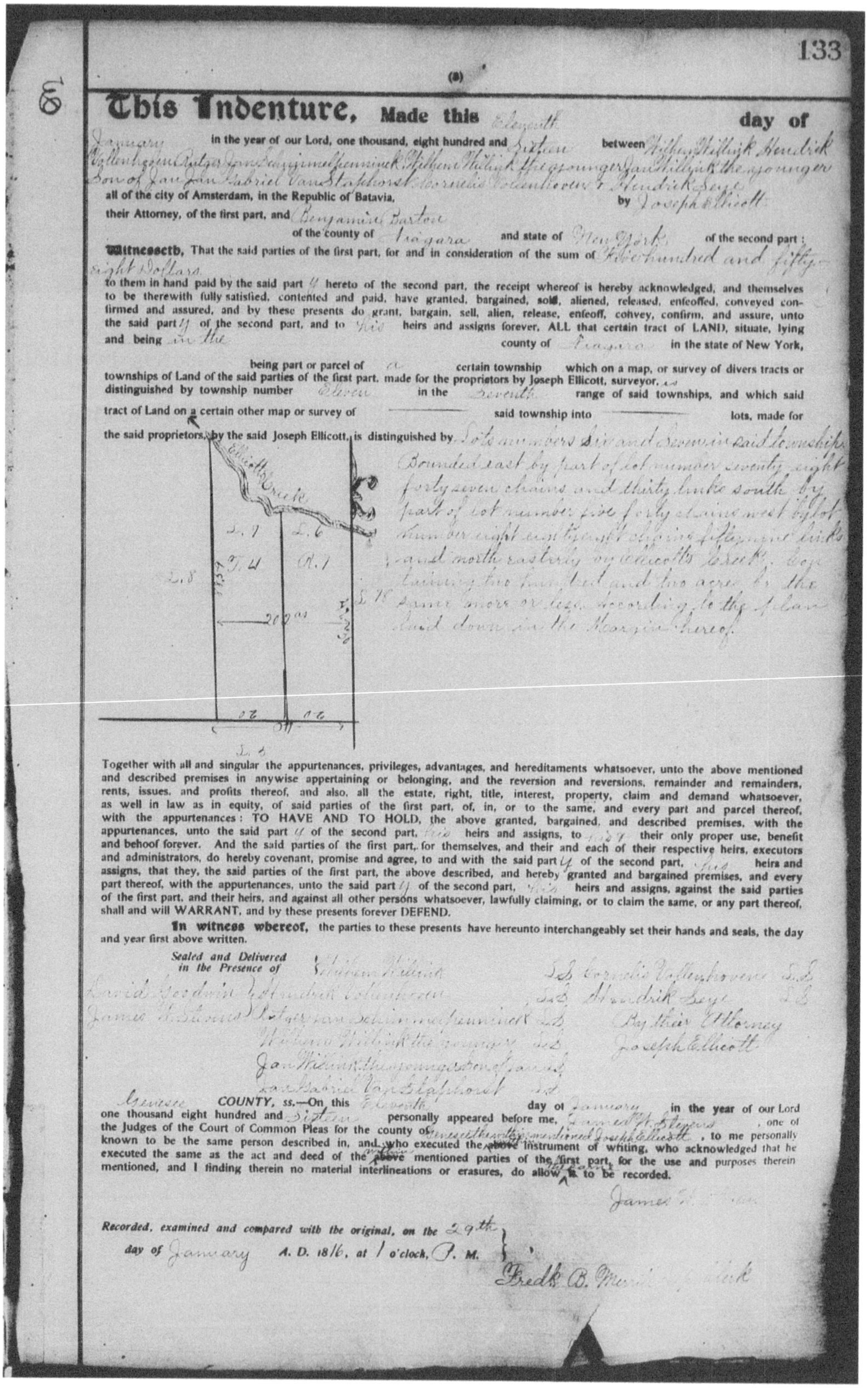

133

(8)

This Indenture, Made this Eleventh **day of** January in the year of our Lord, one thousand, eight hundred and Sixteen between Wilhem Willink Hendrik Vollenhoven Rutger Jan Schimmelpenninck Wilhem Willink the younger Jan Willink the younger son of Jan Jan Gabriel Van Staphorst Cornelis Vollenhoven & Hendrik Seye all of the city of Amsterdam, in the Republic of Batavia, by Joseph Ellicott their Attorney, of the first part, and Benjamin Barton of the county of Niagara and state of New York of the second part:

Witnesseth, That the said parties of the first part, for and in consideration of the sum of Five hundred and fifty-eight Dollars to them in hand paid by the said party hereto of the second part, the receipt whereof is hereby acknowledged, and themselves to be therewith fully satisfied, contented and paid, have granted, bargained, sold, aliened, released, enfeoffed, conveyed confirmed and assured, and by these presents do grant, bargain, sell, alien, release, enfeoff, convey, confirm, and assure, unto the said party of the second part, and to his heirs and assigns forever, ALL that certain tract of LAND, situate, lying and being in the county of Niagara in the state of New York, being part or parcel of a certain township which on a map, or survey of divers tracts or townships of Land of the said parties of the first part, made for the proprietors by Joseph Ellicott, surveyor, is distinguished by township number Eleven in the Seventh range of said townships, and which said tract of Land on a certain other map or survey of ——— said township into ——— lots, made for the said proprietors, by the said Joseph Ellicott, is distinguished by Lots numbers Six and Seven in said township. Bounded east by part of lot number seventy eight forty seven chains and thirty links south by part of lot number five forty chains west by lot number eight ... and north easterly by Ellicotts Creek. Containing two hundred and two acres be the same more or less, according to the plan laid down in the Margin hereof.

Together with all and singular the appurtenances, privileges, advantages, and hereditaments whatsoever, unto the above mentioned and described premises in anywise appertaining or belonging, and the reversion and reversions, remainder and remainders, rents, issues, and profits thereof, and also, all the estate, right, title, interest, property, claim and demand whatsoever, as well in law as in equity, of said parties of the first part, of, in, or to the same, and every part and parcel thereof, with the appurtenances: TO HAVE AND TO HOLD, the above granted, bargained, and described premises, with the appurtenances, unto the said party of the second part, his heirs and assigns, to his their only proper use, benefit and behoof forever. And the said parties of the first part, for themselves, and their and each of their respective heirs, executors and administrators, do hereby covenant, promise and agree, to and with the said party of the second part, his heirs and assigns, that they, the said parties of the first part, the above described, and hereby granted and bargained premises, and every part thereof, with the appurtenances, unto the said party of the second part, his heirs and assigns, against the said parties of the first part, and their heirs, and against all other persons whatsoever, lawfully claiming, or to claim the same, or any part thereof, shall and will WARRANT, and by these presents forever DEFEND.

In witness whereof, the parties to these presents have hereunto interchangeably set their hands and seals, the day and year first above written.

Sealed and Delivered in the Presence of

David Goodwin
James H. Stevens

Wilhem Willink L.S.
Hendrik Vollenhoven L.S.
Rutger Jan Schimmelpenninck L.S.
Wilhem Willink the younger L.S.
Jan Willink the younger son of Jan L.S.
Jan Gabriel Van Staphorst L.S.
Cornelis Vollenhoven L.S.
Hendrik Seye L.S.
By their Attorney
Joseph Ellicott

Genesee COUNTY, ss.—On this Eleventh day of January in the year of our Lord one thousand eight hundred and Sixteen personally appeared before me, James H. Stevens, one of the Judges of the Court of Common Pleas for the county of Genesee the within mentioned Joseph Ellicott, to me personally known to be the same person described in, and who executed the within instrument of writing, who acknowledged that he executed the same as the act and deed of the within mentioned parties of the first part, for the use and purposes therein mentioned, and I finding therein no material interlineations or erasures, do allow it to be recorded.

James H. ...

Recorded, examined and compared with the original, on the 29th *day of* January *A. D. 1816, at* 1 *o'clock,* P. *M.*

Fredk B. Merrill Clerk

Figure 1: Benjamin Barton Deed (Liber 3 p. 133) January 11, 1816

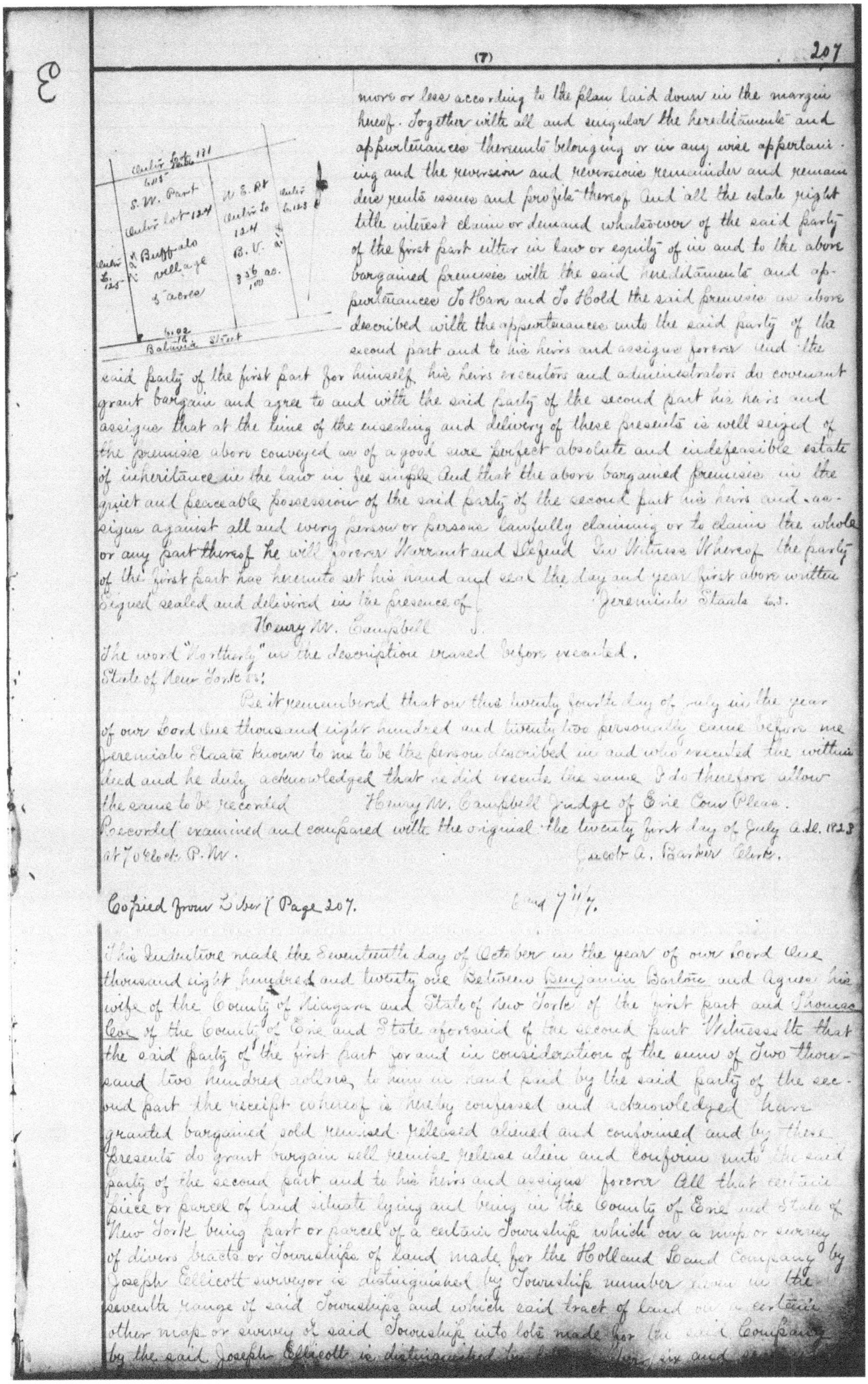

E (7) 207

more or less according to the plan laid down in the margin hereof. Together with all and singular the hereditaments and appurtenances thereunto belonging or in any wise appertaining and the reversion and reversions remainder and remainders rents issues and profits thereof And all the estate right title interest claim or demand whatsoever of the said party of the first part either in law or equity of in and to the above bargained premises with the said hereditaments and appurtenances To Have and To Hold the said premises as above described with the appurtenances unto the said party of the second part and to his heirs and assigns forever And the said party of the first part for himself his heirs executors and administrators do covenant grant bargain and agree to and with the said party of the second part his heirs and assigns that at the time of the ensealing and delivery of these presents is well seized of the premises above conveyed as of a good sure perfect absolute and indefeasible estate of inheritance in the law in fee simple And that the above bargained premises in the quiet and peaceable possession of the said party of the second part his heirs and assigns against all and every person or persons lawfully claiming or to claim the whole or any part thereof he will forever Warrant and Defend In Witness Whereof the party of the first part has hereunto set his hand and seal the day and year first above written

Signed sealed and delivered in the presence of } Jeremiah Staats L.S.
Henry M. Campbell

The word "Northerly" in the description erased before executed.

State of New York ss:

Be it remembered that on this twenty fourth day of July in the year of our Lord One thousand eight hundred and twenty two personally came before me Jeremiah Staats known to me to be the person described in and who executed the within deed and he duly acknowledged that he did execute the same I do therefore allow the same to be recorded Henry M. Campbell Judge of Erie Com Pleas.

Recorded examined and compared with the original the twenty first day of July A.D. 1823 at 7 o'clock P.M. Jacob A. Barker Clerk.

Copied from Liber 7 Page 207. Cmd 7/7.

This Indenture made the Seventeenth day of October in the year of our Lord One thousand eight hundred and twenty one Between Benjamin Barton and Agnes his wife of the County of Niagara and State of New York of the first part and Thomas Coe of the County of Erie and State aforesaid of the second part Witnesseth that the said party of the first part for and in consideration of the sum of Two thousand two hundred dollars to him in hand paid by the said party of the second part the receipt whereof is hereby confessed and acknowledged have granted bargained sold remised released aliened and conformed and by these presents do grant bargain sell remise release alien and conform unto the said party of the second part and to his heirs and assigns forever All that certain piece or parcel of land situate lying and being in the County of Erie and State of New York being part or parcel of a certain Township which on a map or survey of divers tracts or Townships of land made for the Holland Land Company by Joseph Ellicott surveyor is distinguished by Township number seven in the seventh range of said Townships and which said tract of land on a certain other map or survey of said Township into lots made for the said Company by the said Joseph Ellicott is distinguished by [illegible] six and [illegible]

Figure 2: October 17th, 1821 Barton Sells Property Containing the Cemetery to Thomas Coe. (Liber 7 p .207) Page 1

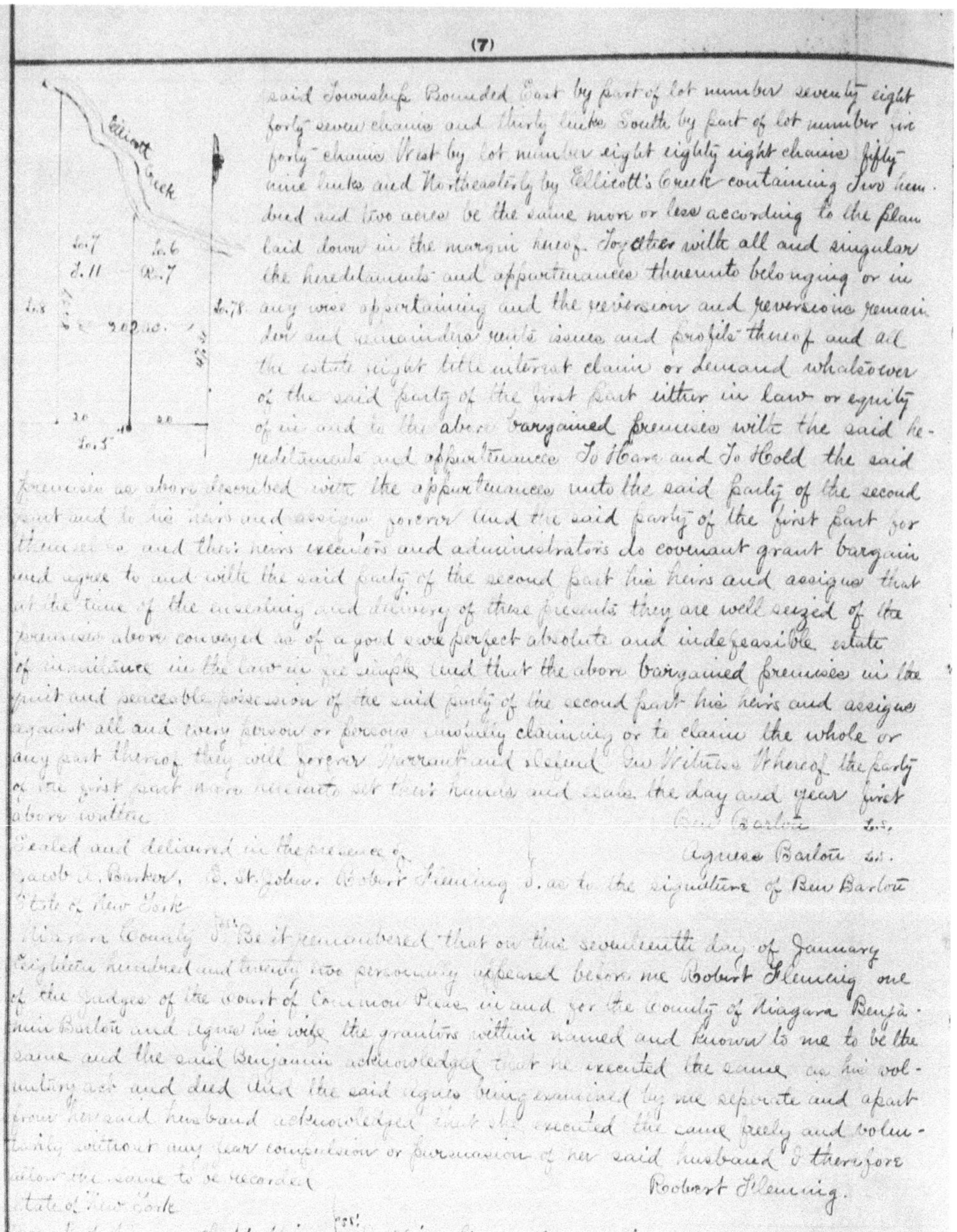

(7)

Lo.7 Lo.6
T.11 R.7
Lo.8 Lo.78
20 20
Lo.5
Ellicott Creek

said Township Bounded East by part of lot number seventy eight forty seven chains and thirty links South by part of lot number six forty chains West by lot number eight eighty eight chains fifty nine links and Northeasterly by Ellicott's Creek containing two hundred and two acres be the same more or less according to the plan laid down in the margin hereof. Together with all and singular the hereditaments and appurtenances thereunto belonging or in any wise appertaining and the reversion and reversions remainder and remainders rents issues and profits thereof and all the estate right title interest claim or demand whatsoever of the said party of the first part either in law or equity of in and to the above bargained premises with the said hereditaments and appurtenances To Have and To Hold the said premises as above described with the appurtenances unto the said party of the second part and to his heirs and assigns forever And the said party of the first part for themselves and their heirs executors and administrators do covenant grant bargain and agree to and with the said party of the second part his heirs and assigns that at the time of the ensealing and delivery of these presents they are well seized of the premises above conveyed as of a good sure perfect absolute and indefeasible estate of inheritance in the law in fee simple and that the above bargained premises in the quiet and peaceable possession of the said party of the second part his heirs and assigns against all and every person or persons lawfully claiming or to claim the whole or any part thereof they will forever Warrant and defend In Witness Whereof the party of the first part have hereunto set their hands and seals the day and year first above written

Ben Barton L.S.
Agnes Barton L.S.

Sealed and delivered in the presence of
Jacob A. Barker, B. St. John. Robert Fleming J. as to the signature of Ben Barton

State of New York
Niagara County ss. Be it remembered that on the seventeenth day of January eighteen hundred and twenty two personally appeared before me Robert Fleming one of the Judges of the Court of Common Pleas in and for the County of Niagara Benjamin Barton and Agnes his wife the grantors within named and known to me to be the same and the said Benjamin acknowledged that he executed the same as his voluntary act and deed and the said Agnes being examined by me separate and apart from her said husband acknowledged that she executed the same freely and voluntarily without any fear compulsion or persuasion of her said husband I therefore allow the same to be recorded

Robert Fleming.

State of New York
County of Niagara clerk's office ss. I Oliver Grace clerk of the said County do certify that Robert Fleming was at the time of taking the acknowledgment of the annexed deed a Judge of the Court of Common Pleas for the said County of Niagara (resident therein) and duly authorized to take the same and that I am acquainted with the hand writing of the said Judge and verily believe the signature to the said certificate of acknowledgment to be his proper hand writing In Testimony Whereof I have hereunto set my hand and affixed the seal of the said County the 18th day of January A.D. 1822. Oliver Grace Clerk.

Recorded examined and compared with the original the twenty second day of July A.D. 1823. at 10 o'clock A.M. Jacob A. Barker Clerk.

Figure 3: October 17th, 1821 Barton Sells Property Containing the Cemetery to Thomas Coe. (Liber 7 p .207) Page 2

459

WARRANTY DEED—COVENANT OF SEIZING (172) Adams & White Company, Binders

This Indenture, Made the fourth day of July in the year of our Lord one thousand eight hundred and fifty one Between William Coe and Phebe Coe his wife of the Town of Cheektowaga in the County of Erie and State of New York. of the first part, and

John Haskell of the Town of Amherst in the County of Erie and State of New York. of the second part.

Witnesseth, That the said party of the first part, for and in consideration of the sum of one thousand dollars.

lawful money of the United States of America, to them in hand paid by the said party of the second part, the receipt whereof is hereby confessed and acknowledged, ha ve granted, bargained, sold, remised, released, aliened, conveyed and confirmed, and by these presents do grant, bargain, sell, remise, release, alien, convey and confirm unto the said party of the second part, and to his heirs and assigns forever.

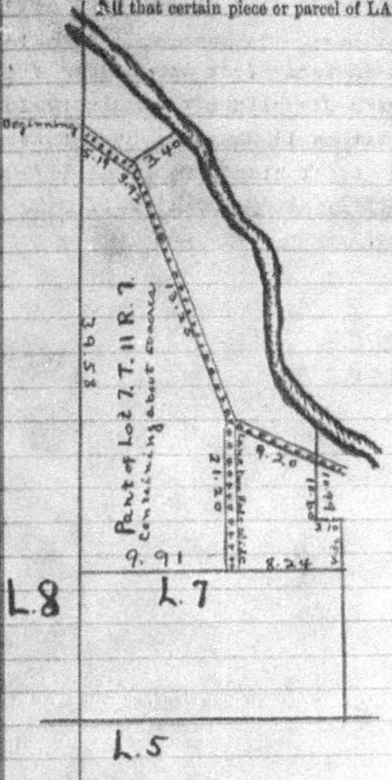

All that certain piece or parcel of LAND, Situate in the Town of Cheektowaga, Erie County and State of New York being a part of Lot Number seven (7) in the eleventh (11th) Township and 7th range according to the Holland Land Company's Survey bounded and described as follows viz: Beginning at a point on the west line of said Lot No. 7 at the distance of about eighty chains and fifty links north of the southwest corner of said lot and in the center of the highway leading southeasterly through said lot; thence southerly along the west line of said Lot No. 7 thirty nine chains and fifty eight links to a post; thence easterly at a right angle with said west line and along the center of a rail fence eighteen chains and fifteen links to a post; thence northerly parallel to the said west line 4 chains and 30 links; thence westerly parallel to the south bounds 2 chains and 10 links to a post; thence northerly bounding on lands owned by the Heirs of Jacob Hershey and others 10 chains 99 links to the Eleven Mile Creek; thence down said Creek at the northwest corner of said Lot No. 7; thence southerly along the west line to the place of beginning, excepting from this conveyance a part of the same lot heretofore conveyed by John Coe to John Haskell hereby conveying about fifty acres be it more or less, according to the within plan, reference being thereto had.

Together with all and singular the tenements, hereditaments and appurtenances thereunto belonging, or in anywise appertaining, and the reversion and reversions, remainder and remainders, rents, issues and profits thereof; and all the estate, right, title, interest, dower and right of dower, claim and demand, whatsoever of the said party of the first part, either in Law or Equity, of, in, and to the above bargained premises, with the said hereditaments and appurtenances.

To Have and to Hold, the said premises as above described, with the appurtenances, unto the said party of the second part, and to his heirs and assigns forever. And the said William Coe for himself and for his heirs, executors and administrators, does covenant, grant, bargain, and agree, to and with the said party of the second part, his heirs and assigns that at the time of the ensealing and delivery of these presents, he was well seized of the premises above conveyed, as of a good, sure, perfect, absolute and indefeasible estate of inheritance in the Law in Fee Simple. And that the above bargained premises, in the quiet and peaceable possession of the said party of the second part his heirs and assigns, against all and every person or persons lawfully claiming or to claim the whole or any part thereof he will forever Warrant and Defend.

In Witness Whereof, the party of the first part ha ve hereunto set their hands and seals the day and year first above written.

Signed, Sealed and Delivered in Presence of

William Coe L.S.

Phebe Ann Coe L.S.

STATE OF NEW YORK, Erie County. } ss.

On this Fifteenth day of May 1855, before me the subscriber ~~personally~~ appeared William Coe and Phebe Ann his wife, and acknowledged that they had severally executed the within instrument. And the said Phebe Ann on a private examination, apart from her husband, acknowledged that she executed the within instrument freely, and without any fear or compulsion of her husband. And I further certify, that I know the persons who made the said acknowledgment to be the individuals described in, and who executed the within instrument.

John Hershey,
Justice of the Peace.

Recorded, Examined and Compared with the Original, the 25th day of September 1857, at o'clock P.M. James H. Pratt Clerk.

Figure 4: July 4th, 1851, William Coe Sells Property Containing the Cemetery to John Haskell. (Liber 172 p.459)

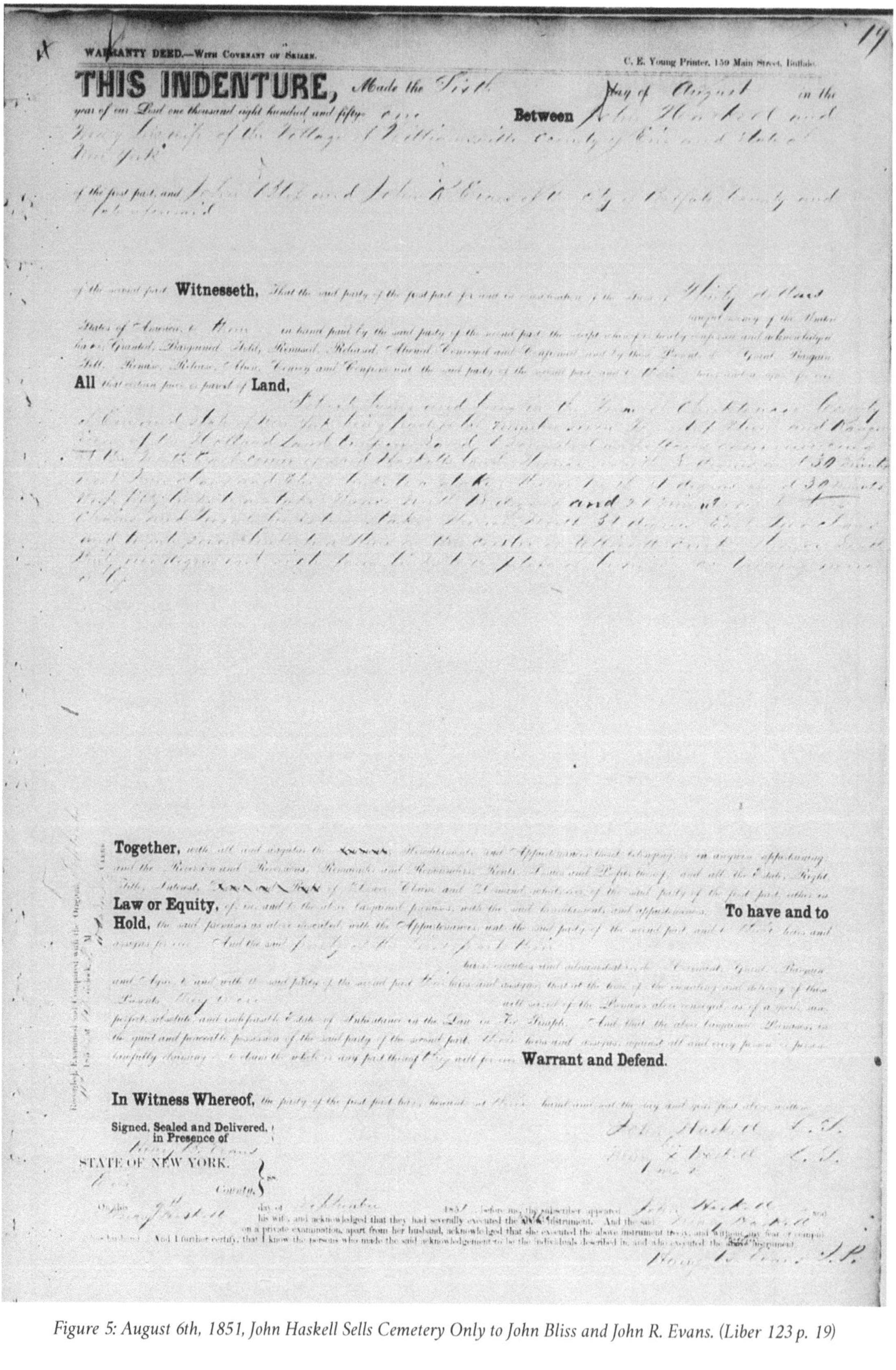
WARRANTY DEED.—With Covenant of Seizin. C. E. Young Printer, 159 Main Street, Buffalo.

THIS INDENTURE, Made the Sixth day of August in the year of our Lord one thousand eight hundred and fifty one **Between** John Haskell and Mary ... wife of the Village of Williamsville County of Erie and State of New York

of the first part, and John Bliss and John R. Evans of the City of Buffalo County and State aforesaid

of the second part **Witnesseth,** That the said party of the first part for and in consideration of the sum of Thirty dollars lawful money of the United States of America, to them in hand paid by the said party of the second part, the receipt whereof is hereby confessed and acknowledged, have Granted, Bargained, Sold, Remised, Released, Aliened, Conveyed and Confirmed, and by these Presents do Grant, Bargain, Sell, Remise, Release, Alien, Convey and Confirm unto the said party of the second part, and to their heirs and assigns for ever, **All** that certain piece or parcel of **Land,**

Together, with all and singular the Hereditaments and Appurtenances thereunto belonging or in anywise appertaining, and the Reversion and Reversions, Remainder and Remainders, Rents, Issues and Profits thereof, and all the Estate, Right, Title, Interest, Claim and Demand whatsoever of the said party of the first part, either in **Law or Equity,** of, in and to the above bargained premises, with the said hereditaments and appurtenances. **To have and to Hold,** the said premises as above described, with the Appurtenances, unto the said party of the second part, and to their heirs and assigns for ever. And the said party of the first part ... for heirs, executors and administrators, do Covenant, Grant, Bargain and Agree to and with the said party of the second part, their heirs and assigns, that at the time of the ensealing and delivery of these Presents they were well seized of the Premises above conveyed, as of a good, sure, perfect, absolute and indefeasible Estate of Inheritance in the Law in Fee Simple. And that the above bargained Premises in the quiet and peaceable possession of the said party of the second part, their heirs and assigns, against all and every person or persons lawfully claiming or to claim the whole or any part thereof they will for ever **Warrant and Defend.**

In Witness Whereof, the party of the first part have hereunto set their hand and seal the day and year first above written.

Signed, Sealed and Delivered, in Presence of

John Haskell L.S.
Mary ... Haskell L.S.

STATE OF NEW YORK, Erie County, ss.

On this 9th day of September 1851 before me, the subscriber appeared John Haskell and Mary Haskell his wife, and acknowledged that they had severally executed the within instrument. And the said Mary Haskell on a private examination, apart from her husband, acknowledged that she executed the above instrument freely, and without any fear or compulsion of her husband. And I further certify, that I know the persons who made the said acknowledgement to be the individuals described in, and who executed the said instrument.

Figure 5: August 6th, 1851, John Haskell Sells Cemetery Only to John Bliss and John R. Evans. (Liber 123 p. 19)

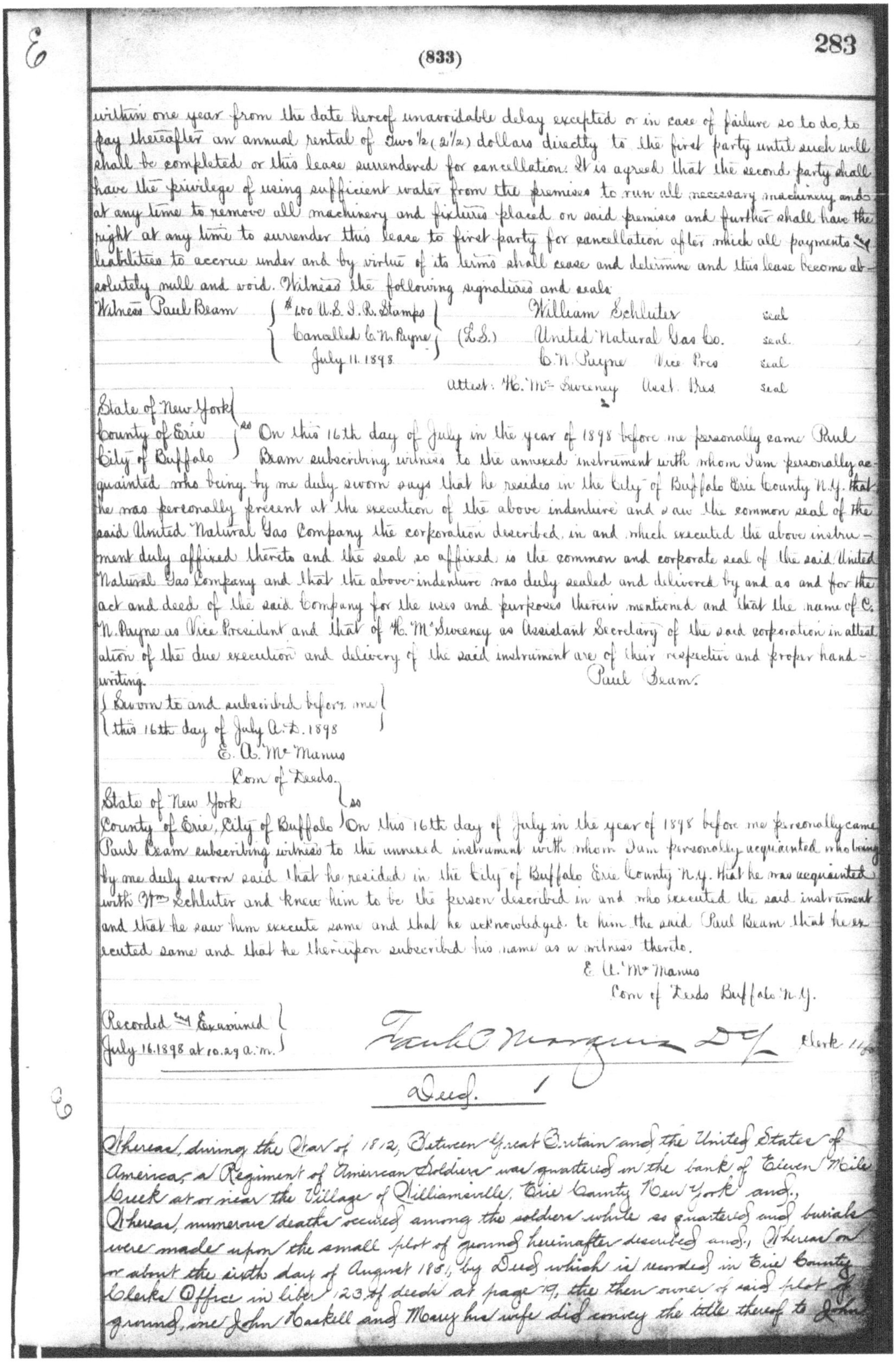
E

(833) 283

within one year from the date hereof unavoidable delay excepted or in case of failure so to do, to pay thereafter an annual rental of Two ½ (2½) dollars directly to the first party until such well shall be completed or this lease surrendered for cancellation. It is agreed that the second party shall have the privilege of using sufficient water from the premises to run all necessary machinery and at any time to remove all machinery and fixtures placed on said premises and further shall have the right at any time to surrender this lease to first party for cancellation after which all payments & liabilities to accrue under and by virtue of its terms shall cease and determine and this lease become absolutely null and void. Witness the following signatures and seals

Witness Paul Beam { #.00 U.S. I.R. Stamps } William Schluter seal
{ Cancelled C.N. Payne } (L.S.) United Natural Gas Co. seal
{ July 11. 1898. } C.N. Payne Vice Pres seal
Attest: H. McSweeney Asst. Secy seal

State of New York }
County of Erie } ss On this 16th day of July in the year of 1898 before me personally came Paul
City of Buffalo } Beam subscribing witness to the annexed instrument with whom I am personally acquainted who being by me duly sworn says that he resides in the City of Buffalo Erie County N.Y. that he was personally present at the execution of the above indenture and saw the common seal of the said United Natural Gas Company the corporation described in and which executed the above instrument duly affixed thereto and the seal so affixed is the common and corporate seal of the said United Natural Gas Company and that the above indenture was duly sealed and delivered by and as and for the act and deed of the said Company for the uses and purposes therein mentioned and that the name of C. N. Payne as Vice President and that of H. McSweeney as Assistant Secretary of the said corporation in attestation of the due execution and delivery of the said instrument are of their respective and proper handwriting.
Paul Beam.

{ Sworn to and subscribed before me }
{ this 16th day of July A.D. 1898 }
E. A. McManus
Com of Deeds.

State of New York } ss
County of Erie, City of Buffalo } On this 16th day of July in the year of 1898 before me personally came Paul Beam subscribing witness to the annexed instrument with whom I am personally acquainted who being by me duly sworn said that he resided in the City of Buffalo Erie County N.Y. that he was acquainted with Wm Schluter and knew him to be the person described in and who executed the said instrument and that he saw him execute same and that he acknowledged to him the said Paul Beam that he executed same and that he thereupon subscribed his name as a witness thereto.
E. A. McManus
Com of Deeds Buffalo N.Y.

Recorded & Examined {
July 16. 1898 at 10.29 a.m. } Frank C. Marquis D.Y. Clerk

Deed. 1

E

Whereas, during the War of 1812, Between Great Britain and the United States of America, a Regiment of American Soldiers was quartered on the bank of Eleven Mile Creek at or near the Village of Williamsville, Erie County New York and,
Whereas, numerous deaths occurred among the soldiers while so quartered and burials were made upon the small plot of ground hereinafter described and, Whereas on or about the sixth day of August 1851, by Deed which is recorded in Erie County Clerks Office in liber 123 of deeds at page 19, the then owner of said plot of ground, one John Haskell and Mary his wife did convey the title thereof to John

Figure 6: July 12th, 1898, the Estates of John Bliss and John R. Evans Deeded the Cemetery (Parcel A) to the Buffalo Historical Society. (Liber 833 p. 283) Page 1

284

(833)

E

Bliss and John R. Evans with the intent and purpose of dedicating said plot of ground perpetually to the memory of said deceased soldiers and, Whereas said John Bliss and John R. Evans have both since died, leaving the undersigned parties of the first part their sole heirs at law and, Whereas, The Buffalo Historical Society of Buffalo, Erie County, New York is willing to take the title to said burial ground and to give the same proper care and attention, and in the very nature of things appears to be the Ideal custodian for such a little burial ground associated as it is with the history of Erie County. Now therefor, with the sole intent and purpose of dedicating said premises for the sole purpose they are now used and to place the record title thereof in a suitable custodian where the same will be kept perpetually sacred to the memory of said deceased soldiers of the War of 1812,

We the undersigned Rachel L. Evans, Francis T. Evans, Miriam H. Evans, John T. Evans and Georgiana H. Evans his wife and Martha C. Evans, sole surviving children and heirs at law of John R. Evans, late of Erie County N.Y. now deceased and John Horace Bliss of Erie Pennsylvania sole surviving heir at law of John Bliss late of Erie County N.Y. now deceased for a good and valuable consideration to them in hand paid the receipt of which is hereby acknowledged have bargained, sold remised and quit claimed and by these presents do bargain, sell, remise, and quit claim unto the Buffalo Historical Society of Buffalo, Erie County New York and its successors forever and in consideration of one Dollar ($1.00).

All that certain tract or parcel of land situate lying and being in the Town of Cheektowaga, County of Erie and State of New York and being part of lot number seven (7) of the Holland Land Company's land and bounded as follows; Commencing at the northeast corner of said Haskell's land, thence south eight (8) degrees and thirty (30) minutes west four (4) chains and eleven (11) links to a stake, thence north eighty one (81) degrees thirty (30) minutes west fifty (50) links to a stake, thence north eighteen (18) degrees and twenty (20) minutes west three (3) chains and twenty (20) links to a stake, thence north fifty (50) degrees east two (2) chains and twenty seven (27) links to a stone in the center of Ellicott Creek, thence south thirty five (35) degrees east sixty four (64) links to the place of beginning, containing more or less together with the appurtenances and all the estate and rights of the said parties of the first part, the said heirs of John Bliss and John R. Evans deceased in and to the premises above described. To Have and to Hold the above bargained premises unto the said Buffalo Historical Society and its successors forever said premises to remain and be perpetually a cemetary and resting place for the dead soldiers of the War of 1812.

In Witness Whereof, the parties of the first part have hereunto set their hands and seals this 12th day of July, A.D. 1898.

Rachel L. Evans (L.S.)
Miriam Hunt Evans (L.S.)
John T. Evans (L.S.)
Georgiana H. Evans (L.S.)
Martha C. Evans (L.S.)
Jno. H. Bliss (L.S.)

State of New York
County of Erie } ss
City of Buffalo

On this 12th day of July 1898 before me the subscriber personally came Rachel L. Evans, Miriam Hunt Evans, John T. Evans, Georgiana H. Evans, Martha C. Evans to me known to be the same persons named in and who executed the foregoing instrument and they severally

Figure 7: July 12th, 1898, the Estates of John Bliss and John R. Evans Deeded the Cemetery (Parcel A) to the Buffalo Historical Society. (Liber 833 p. 283) Page 2

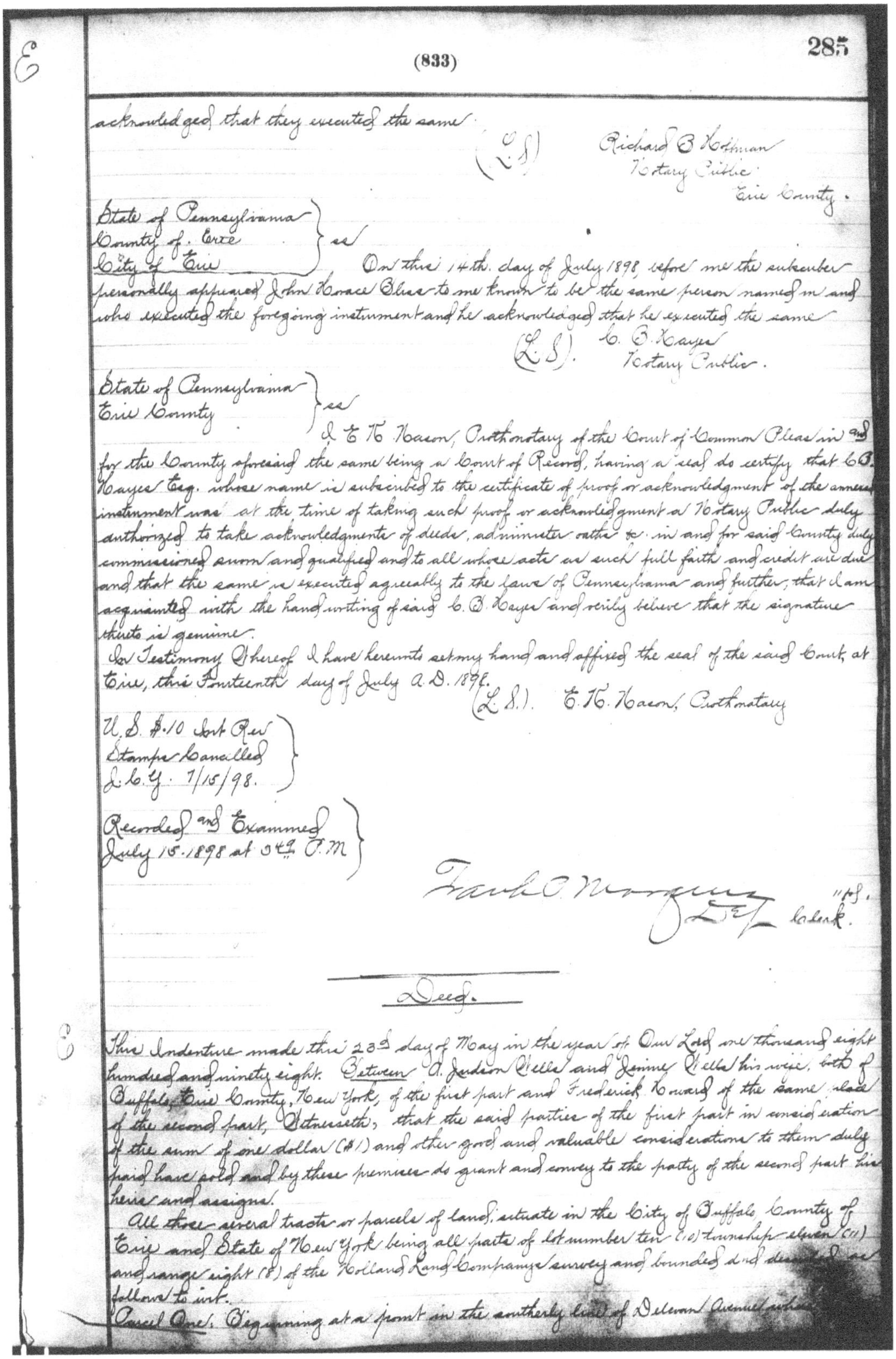

E

(833) 285

acknowledged that they executed the same. (L.S.) Richard B Hoffman
Notary Public
Erie County.

State of Pennsylvania }
County of Erie } ss
City of Erie }
On this 14th day of July 1898, before me the subscriber personally appeared John Horace Bliss to me known to be the same person named in and who executed the foregoing instrument and he acknowledged that he executed the same
(L.S.) C. B. Hayes
Notary Public.

State of Pennsylvania }
Erie County } ss
I E. H. Nason, Prothonotary of the Court of Common Pleas in and for the County aforesaid the same being a Court of Record, having a seal do certify that C.B. Hayes Esq. whose name is subscribed to the certificate of proof or acknowledgment of the annexed instrument was at the time of taking such proof or acknowledgment a Notary Public duly authorized to take acknowledgments of deeds, administer oaths &c. in and for said County duly commissioned sworn and qualified and to all whose acts as such full faith and credit are due and that the same is executed agreeably to the laws of Pennsylvania and further, that I am acquainted with the handwriting of said C. B. Hayes and verily believe that the signature thereto is genuine.
In Testimony Whereof I have hereunto set my hand and affixed the seal of the said Court at Erie, this Fourteenth day of July A.D. 1898.
(L.S.) E. H. Nason, Prothonotary

U.S. $.10 Int Rev
Stamps Cancelled
J.C.Y. 7/15/98.

Recorded and Examined
July 15. 1898 at 0 4? P.M.

[illegible signature] Clerk.

Deed.

E

This Indenture made this 23d day of May in the year of Our Lord one thousand eight hundred and ninety eight. Between O. Judson Wells and Jennie Wells his wife, both of Buffalo, Erie County, New York, of the first part and Frederick Howard of the same place of the second part, Witnesseth, that the said parties of the first part in consideration of the sum of one dollar ($1) and other good and valuable considerations to them duly paid have sold and by these premises do grant and convey to the party of the second part his heirs and assigns.
All those several tracts or parcels of land, situate in the City of Buffalo County of Erie and State of New York being all parts of lot number ten (10) township eleven (11) and range eight (8) of the Holland Land Company's survey and bounded and described as follows to wit.
Parcel One. Beginning at a point in the southerly line of Delavan Avenue where

Figure 8: July 12th, 1898, the Estates of John Bliss and John R. Evans Deeded the Cemetery (Parcel A) to the Buffalo Historical Society. (Liber 833 p. 283) Page 3

In an account as described in the *Buffalo Cemeteries: An account of the Burial-places of Buffalo, From the Earliest Times*, read before the Buffalo Historical Society, February 4, 1879:

It would not be right to omit a notice of the soldier's burial place at Williamsville. About six thousand of our army raised during the first year of the War of 1812 -15 and sent on to protect our frontier at Buffalo, went into winter quarters at Williamsville Village, situated eleven miles north of Buffalo. Their encampment at that place was just north of the main road and contiguous to the village on the extensive premises then owned by the enterprising merchant and milling firm of Juba Storrs & Co. This ground continued to be occupied by our soldiers more or less during the war. Sickness, as is usual in camp, prevailed among them; and some two or three hundred died and were buried on the grounds adjacent. Since then, the Village has spread, covering the ground where they were buried and long since not a vestige of a grave was left to be seen. There has been no one to look after or care for their bones when exhumed, as they often were, in excavating cellars or making improvements such as are necessary in a growing village.

There were quite extensive barracks built on the bank of the Eleven Mile Creek, a very healthy, eligible place, it being retired from the immediate scenes of conflict and about one mile above the Village. It continued to be our general hospital for sick and wounded soldiers during this war.

During the three years of the conflict, many of our soldiers died and were buried on the side of a nearby field. The ground that contains their remains is approximately half an acre in size. It lies on the southwesterly side of a public road; the Eleven Mile Creek running parallel and adjoining.

Two of the Village's townsmen, Col. John Bliss and John B. Evans, feeling an interest that this ground should never be disturbed or encroached upon, procured the title by a warranty deed from the owners, John Haskell and his wife, to themselves, dated August 6th, 1851, and duly recorded. They have both since deceased, leaving the title to their heirs. The number of our own soldiers buried on this ground is supposed to be three hundred or more; and in one retired corner lie nearly one hundred of our enemies who were wounded, taken prisoners and died. They captured many of the latter at the desperate sortie the British made on Fort Erie, August 14, 1814, and the explosion of its magazine. There were several hundred prisoners taken at this encounter; many of them were wounded most horribly, having been blown up when the magazine exploded.

On the second day after the sortie, I saw a number of wagon loads of those blackened and maimed British soldiers as they stopped in front of my father's house on their way to the hospital.

Recently, with one of the oldest residents of the town, Hon. T. A. Hopkins, I visited this burying place. There was not a slab or monument of any kind to be seen to designate the graves. Only the uneven hillocks marked the spot that contains the bones of our un-honored dead.

This ground is a little elevated from the surrounding land and road adjoining. On the border of two sides stand ten sugar maple trees, from one and one-fourth to two feet in diameter: all but one in a fine, healthy condition. I have no doubt they were planted there by the comrades of those whose bones now occupy this ground. The tenth tree is in a state of decay, and like many of the human race, is beginning to die at the top.

It is known only to a few living witnesses that this spot of ground contains the graves of some of our dead soldiers. Our government ought to erect at this beautiful, retired place a suitable monument to their memory and provide for removing thither the skeletons of others, when found in excavating in the village.

As you can see from this earliest known photograph of the graveyard, the five sugar maple trees on the eastern border of the graveyard as described in the historical account.

Possibly One of the Oldest Known Photographs of War of 1812 Cemetery

Most likely 1898 or 1899 after the estates of John Bliss and John R. Evans deeded the cemetery to the Buffalo Historical Society. Photo courtesy of David Przepiora.

Cemetery with Cannon and Gate

On 9 November, 1898, the Buffalo Historical Society placed a cannon at the cemetery. The cannon was a Parrott Gun taken from the north end of the circle at The Front and was suitably inscribed in the memory of the dead soldiers. An arched gateway was installed with the letters "Soldiers of the War of 1812" cut from galvanized iron. On August 25th, 1942, the original cannon (acquired in 1898) was presented to the Bethlehem Steel Corporation to be used as scrap metal for the war effort.

As described in the Buffalo Cemeteries: An Account of the Burial-places of Buffalo, from the Earliest Times, read before the Buffalo Historical Society, February 4, 1879:

On the border of two sides stand ten sugar maple trees, from one and one-fourth to two feet in diameter: all but one in a fine, healthy condition. I have no doubt they were planted there by the comrades of those whose bones now occupy this ground. The tenth tree is in a state of decay, and like many of the human race, is beginning to die at the top.

. . .

I believe five sugar maple trees on each side which are shown on the left of the photo.

After carefully looking through the archives for maps and aerial photographs of the area in which the burial ground stood, I could compile a scaled overlay of the area onto Google Earth-.The cemetery marker on the map signifies the location from the canon at the present time. The three remaining sugar maple trees are visible of the original 10 planted and the black dot at the line is possibly the Parrot Gun installed in 1898 by the Buffalo Historical Society.

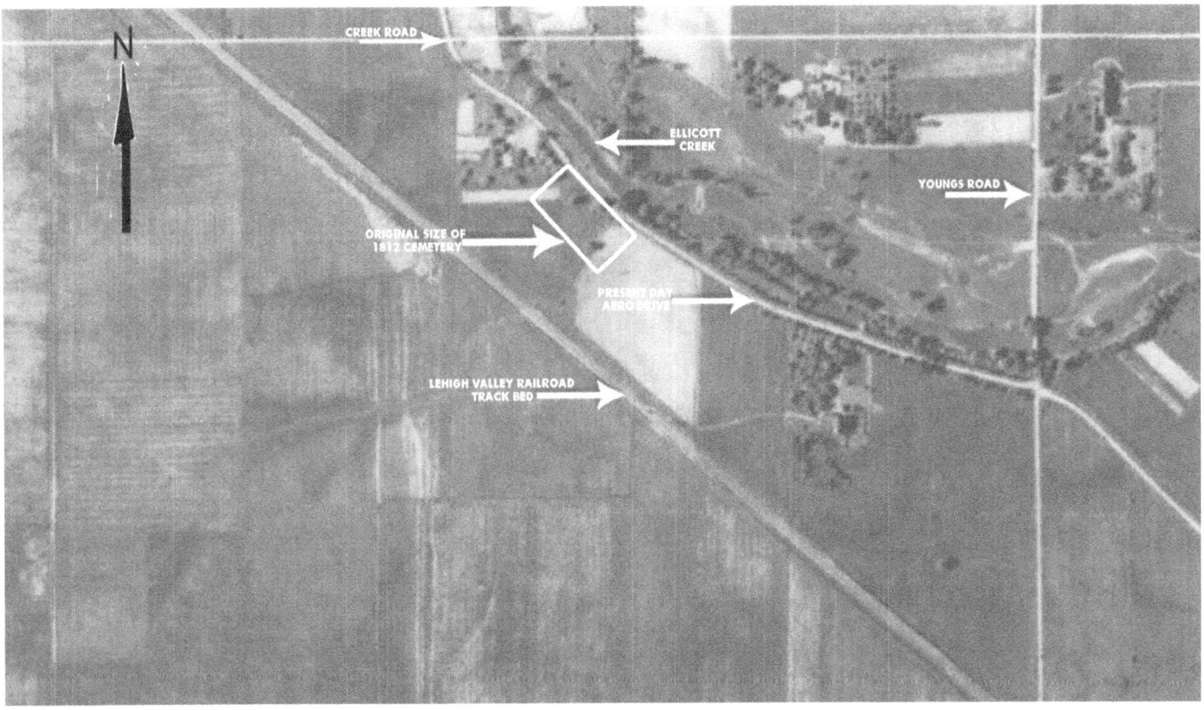

1927 Aerial Photograph of the burial ground

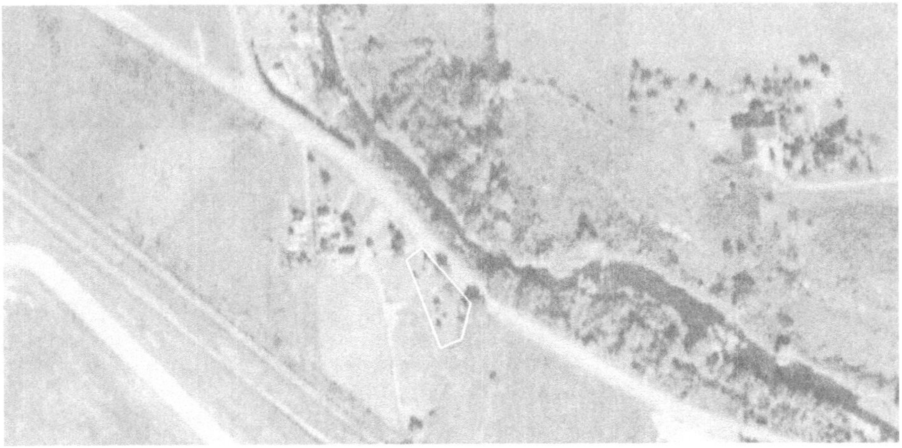

1951 Aerial Photograph of the Graveyard

Present-day satellite image of burial ground lines overlaying the boundary of the original graveyard from 1927 aerial photograph.

Lines Represent Current Day burial ground Boundaries

LiDar burial ground Area

In this image the present-day burial ground boundaries are marked and the original burial ground boundaries outlined. As one can see, a good portion of the original burial ground is now a driveway and parking area.

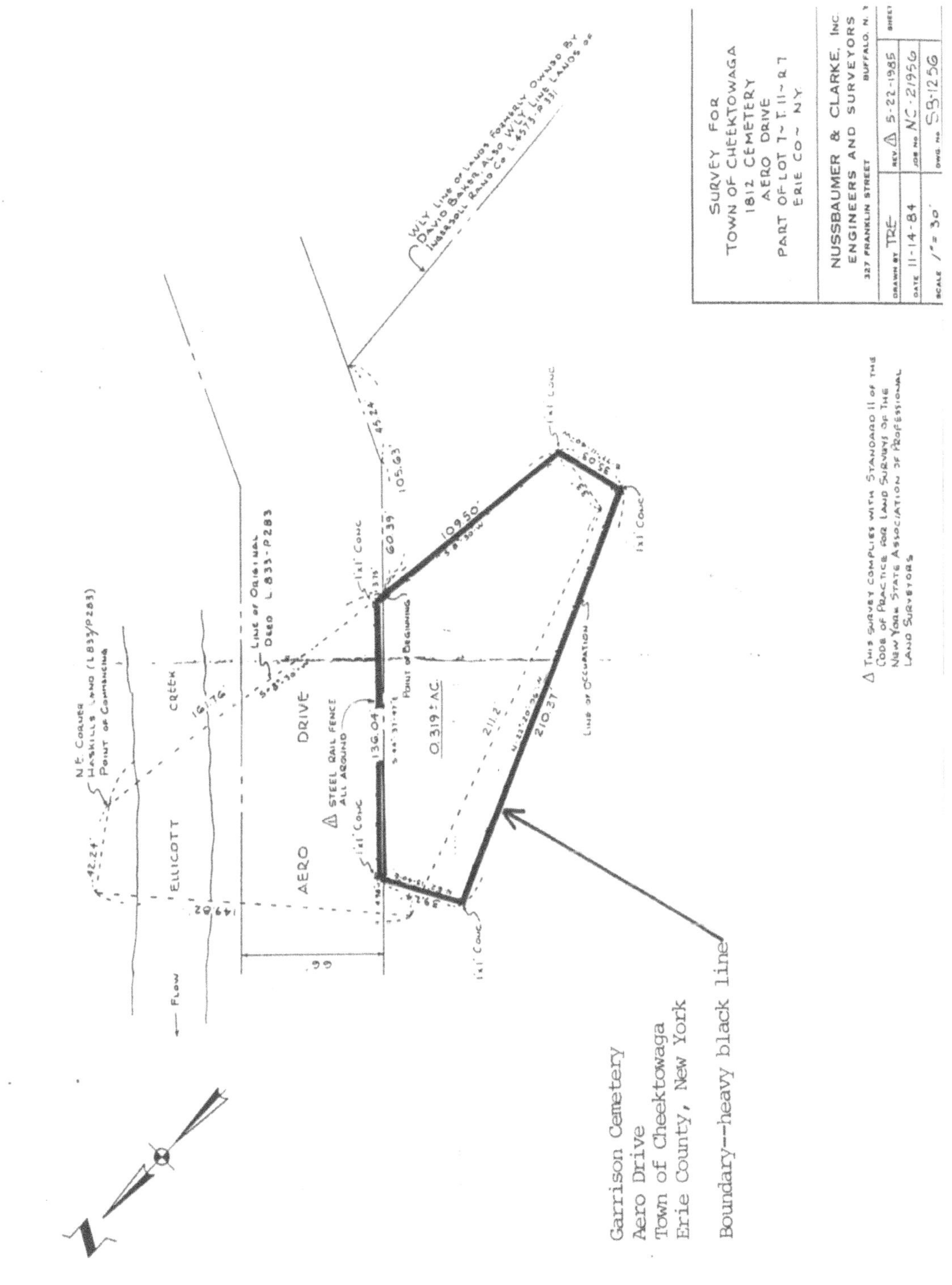

Survey of War of 1812 Cemetery

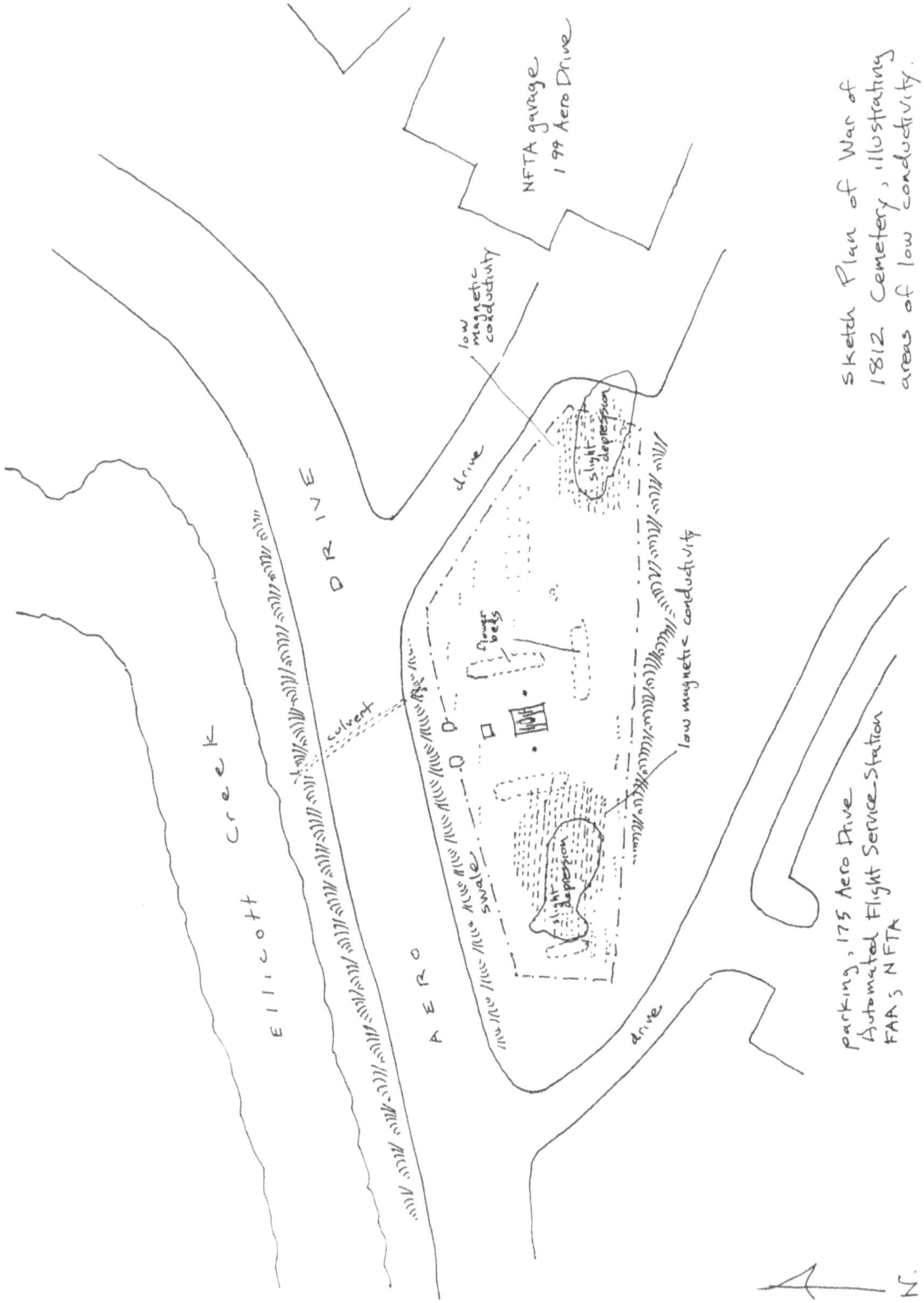

Sketch Plan of War of 1812 Cemetery

From Niagara Frontier #12 1965 pp. 51-57, *Jacob Porter Norton, A Yankee on the Niagara Frontier:*

Jacob Norton was a young lieutenant assigned to the 4th Infantry Regiment during the War of 1812. During his time of service, he wrote letters home to his father and kept a diary of his activity. Among his entries in his diary are; November 14th, 1814 "The 4th Regt marched from our encampment near Buffalo, and took quarters in the Hospital Barracks, at Eleven Mile Creek in Williamsville, 11 miles from Buffalo." June 1815, while writing his thoughts about fellow officers, he had this to say about Lt Col Ranney, "The army burying ground at this place when we first came, was in a miserable situation, - Many of the graves were not more than half filled, and they had been dug promiscuously without regard to order, so that it was difficult in some places to distinguish the graves. Lt Col Ranney had them mounded up and clotted with sard, the stumps dug up, a fence built round it, a gate with an arch over the way, upon the arch was inscribed, "Sacred to the memory of those men who died of wounds received in the memorable fields of Chippewa, Bridgewater and Erie, - They rest in honor and deserve the gratitude of their country."

On June 14th, 1815, the 4th Infantry Regt marched out of Williamsville beginning their journey home.

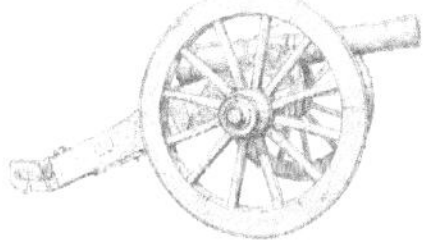

7

LIST OF BURIED U.S. SOLDIERS AT THE GARRISON BURIAL GROUND

In 1813, the local barracks underwent expansion to serve as a hospital with Dr. Mann as the attending surgeon. The facility received and treated over eleven hundred sick and wounded soldiers, including some British prisoners transferred from Lewiston, New York. This event transformed the entire area into a vital hospital base.

Soldiers who succumbed to their injuries during their treatment at this military hospital were laid to rest in a cemetery situated on Aero Drive, then within the boundaries of Amherst. The construction of the New York State Thruway nearly 150 years later necessitated the relocation of the cemetery, which now falls under the jurisdiction of Cheektowaga.

A noteworthy observation is that of the 432 soldiers interred in this cemetery, only twelve hailed from New York State. The remaining soldiers, comprising infantrymen and artillerymen, represented a diverse range of states, including New Hampshire, Vermont, Massachusetts, Connecticut, New Jersey, Pennsylvania, Maryland, Virginia, North Carolina, Ohio, and Kentucky. The database of names were compiled by Cheektowaga Town Historian Maureen E. Gleason and David Przepior.

Name	DoD	Unit	Commanding Officer
Calmon James W	29 Aug 1814	1st US Inf	Capts Symmes & Clemson
Caulker Michael	24 Aug 1814	1st US Inf	Capts Symmes & Clemson
Conker Michael	23 Aug 1814	1st US Inf	Capt John C Symmes
Coulker Michael	24 Aug 1814	1st US Inf	Capt H Stark
Critichlow John	1 Sept 1814	1st US Inf	Capt John C Symmes
Detwiller David	15 Dec 1814	1st US Inf	Capts Hamilton & Owens
Emons John	12 Jan 1815	1st US Inf	Capts Johnson & Symmes
Fegan William	25 Dec 1814	1st US Inf	Capts Stark & Owens
Henderson James	28 Oct 1814	1st US Inf	Capts Symmes & Clemson
Hill Samuel H	1 Jan 1814	1st US Inf	Capt F Hamilton
Kenney Daniel Corporal	11 Sept 1814	1st US Inf	Capt John C Symmes
Kinkaid James	1 Dec 1814	1st US Inf	Capt S Owens
Mc Colman/Mc Calmon James	29 Aug 1814	1st US Inf	Capt John C Symmes
Mc Mullen John	16 Oct 1814	1st US Inf	Capt Simon Owens
Mc Neal Samuel	1 Feb 1815	1st US Inf	Capt John C Symmes
Mellen Jacob	28 Oct 1814	1st US Inf	Capt Hamilton
Miller Jacob	28 Nov 1814	1st US Inf	Capts Clemson & Symmes
Mullen John	18 Oct 1814	1st US Inf	Capt Hamilton
Neal William	19 Jan 1815	1st US Inf	Capt John C Symmes
TenEykck Samuel	11 Sept 1814	1st US Inf	Capt John C Symmes
Morris Joseph	27 Dec 1814	1st US Rifles	Capt Thomas Ramsey
Price Franklin Corporal	11 Dec 1814	1st US Rifles	Capt Thomas Ramsey
Souther Samuel	1/13 Jan 1815	1st US Rifles	Capt Edward Wadsworth
Bridges Daniel	7 Nov 1814	4th US Inf	Capt Josiah Bacon
Chase Benjamin	11 Feb 1815	4th US Inf	Capt John Smith
Colby Nicholas	24 Feb 1815	4th US Inf	Capt Josiah Bacon
Littlefield John	11 Jan 1815	4th US Inf	Capt A Hawkins
Roundy Ralph	10 Feb 1815	4th US Inf	Capt Frederick Conklin
Silver John	25 Jan 1815	4th US Inf	CaptL ewis Peckham
Simonds Sylvester	24 Jan 1815	4th US Inf	Capt Fred Conkling
Sylver John	30 Jan 1815	4th US Inf	Unknown
Trask Nehemiah	17 Feb 1815	4th US Inf	Capt George Gooding
Watts William	18 Nov 1814	4th US Inf	Capt Larabee
Stewart Andrew	16 Dec 1814	4th US Rifles	Capt Joseph Kean
Conlon James	7 Dec 1814	5th US Inf	Capts Dorman & Jamison
Fisher John	3 Nov 1815	5th US Inf	Capt James Dorman
Nagle Christian	17 Apr 1813	5th US Inf	Capt Richard Whartenby
Pryor James	13 Jan 1815	5th US Inf	Capt Leroy Opie
Shaw Thomas	12 Dec 1814	5th US Inf	Capt James Dorman
Andrews Seth	Unknown	6th US Inf	Unknown
Allen Noah Corporal	1 Nov 1814	9th US Inf	Capt Turner Crooker
Chase Elkanah	25 July 1814	9th US Inf	Capt Georgr Bender Jr
Crooker Calvin	16 Oct 1814	9th US Inf	Lt Daniel Chandler
Ditson Joseph	Unknown	9th US Inf	Capt David S Townsend
Eldridge Levi	July/Aug 1814	9th US Inf	Capt Ropes
Elliot James	4 Jan 1814	9th US Inf	Capt Turner Crooker

Fales/Fails Plina	28 Jan 1814	9th US Inf	Lt W L Foster
Hadley Amos	6 Aug 1814	9th US Inf	Capt Thomas J Harrison
Harrington Isaac	28 Dec 1814	9th US Inf	Lt Daniel Chandler
Hart Charles	11 Jan 1814	9th US Inf	Capt D S Townsend
Hart James	20 Nov 1814	9th US Inf	Capt Thomas J Harrison
Heard Timothy	Nov 1813	9th US Inf	Capt Mose Holt
Horsham Joel	Unknown	9th US Inf	Capt Ebenezer Childs
Johnson Jesse	2 Jan 1814	9th US Inf	Capt David S Townsend
Johnson William	12 Sept 1814	9th US Inf	Capt Thomas J Harrison
Keeler Isaac	25 Jan 1815	9th US Inf	Capt Samuel L Allen
Kelph Peter	15 Jan 1814	9th US Inf	Capt J F Norris
Lord Warren	30 Nov 1813	9th US Inf	Capt Ebenezer Childs
Luce Noah	Unknown	9th US Inf	Capt Chester Lyman
Manning Nathaniel Corporal	16 Sep 1814	9th US Inf	Lt Daniel Chandler
Marshall James	22 Jan 1814	9th US Inf	Capt D S Townsend
Mc Lahlan Hugh	Unknown	9th US Inf	Capt Ebenezer Childs
Morey James	Unknown	9th US Inf	Lt Samuel L Allen
Morse Nathaniel	17 Jan 1815	9th US Inf	Capt Ebenezer Childs
Mory James	26 Dec 1814	9th US Inf	Lt Jared Ingersoll
Reed Samuel	27 Oct 1814	9th US Inf	Capt Samuel L Allen
Sealy Ruben	15 Apr 1814	9th US Inf	Capt Norris
Skelton William	20 Nov 1814	9th US Inf	Capt Ebenezer Childs
Whipple William	3 Jun 1815	9th US Inf	Lt Jared Ingersoll
Wyer David	15 Aug 1814	9th US Inf	Capt Thomas J Harrison
Booth Gray	25 Jan 1815	10th US Inf	Capt Joseph Clay
Brooks Eli	30 Jan 1815	10th US Inf	Capt Robert Mitchell
Chesnutt Duke W	15 Feb 1815	10th US Inf	Capt James Wellborn
Danson John	25 Jan 1815	10th US Inf	Capt Joseph Clay
Dawson John	25 Jan 1815	10th US Inf	Capt Joseph Clay
Freet George	17 Feb 1815	10th US Inf	Capt Joseph Clay
Patterson William	15 Jan 1815	10th US Inf	Capt Joseph Clay
Stallings Hardy	1 Feb 1815	10th US Inf	Capt Joseph Clay
Taylor John Sergeant	Unknown	10th US Inf	Capts Vashon & Bailey & Nelson
Taylor Redding	20 Mar 1815	10th US Inf	Capt Robert Mitchell
Tyler Redding	16 Feb 1815	10th US Inf	Capt Robert Mitchell
Wallace John	31 Nov 1814	10th US Inf	Capt Bryant & Nelson
Warick James	22 Dec 1814	10th US Inf	Capt George Vashon
Alvord Elisha	11 Nov 1814	11th US Inf	Capt William S Foster
Atwood Elisha	11 Nov 1814	11th US Inf	Capt William S Foster
Badger Isaac	18 Sept 1814	11th US Inf	Capt William S Foster
Belding Jonathan	25 June 1815	11th US Inf	Lt Isaac Clark
Bradford Timothy	15 Apr 1814	11th US Inf	Capt Valentine Goodrich
Brown Ephraim	13 Aug 1814	11th US Inf	Lt Isaac Clark
Cady Aden	9 Aug 1814	11th US Inf	Capt Samuel H Holley
Campbell James B Colonel	4 Sept 1814	11th US Inf	Unknown
Doolittle Abraham	22 Aug 1814	11th US Inf	Capt B Smead
Ferris Gideon P	24 Aug 1814	11th US Inf	Capt John Bliss

Fry Asa	14 Sept 1814	11th US Inf	Capt William S Foster
Fry Isaac	14 Sept 1814	11th US Inf	Unknown
Goodenow Hezekiah	10 Feb 1815	11th US Inf	Capt William S Foster
Gould John Jr.	30 Oct 1814	11th US Inf	Capt William S Foster
Graves Waine	18 Nov 1814	11th US Inf	Capt William S Foster
Grimes James	19 Sept 1814	11th US Inf	Capt William S Foster
Hall Lyman	26 Oct 1814	11th US Inf	Capt William S Foster
Harris John	26 Aug 1814	11th US Inf	Capt Bliss & Smead
Hildreth Peter	5 Aug 1814	11th US Inf	Capt John Bliss
Hill William	30 Aug 1814	11th US Inf	Capt Valentine R Goodrich
Hull Lyman	1814	11th US Inf	Capt John W Weeks
Lainhard David	25 Aug 1814	11th US Inf	Capt Malachi Corning
Lock James	Unknown	11th US Inf	Capt William S Foster
Lovering Jonathan	1 Nov 1814	11th US Inf	Capt William S Foster
Martin David	30 Aug 1814	11th US Inf	Unknown
Mc Pherson John	Unknown	11th US Inf	Capt John W Weeks
Moulton Richard	25 Sep 1814	11th US Inf	Capt William S Foster
Nichols John	30 Nov 1814	11th US Inf	Capt William S Foster
Parkhurst Thomas	31 Dec 1814	11th US Inf	Capt William S Foster
Peck Henry	27 Sep 1814	11th US Inf	Capt William S Foster
Ross John	22 Sept 1814	11th US Inf	Lt B Smead
Severance Jonathan	12 Nov 1814	11th US Inf	Capt William S Foster
Sovereign Jonathan	1 Nov 1814	11th US Inf	Capt William S Foster
Sprague Eliakiam	20 Oct 1814	11th US Inf	Unknown
Stephens James	30 Jan 1815	11th US Inf	Capt William S Foster
Warren John Corporal	3 Oct 1814	11th US Inf	Capt William S Foster
White John	2 Nov 1814	11th US Inf	Capts Hale & Foster
Whitman John	8 Nov 1814	11th US Inf	Capt William S Foster
Bell Thomas	Jan 1815	12th US Inf	Capt James Paxton
Bowman Adam	28 Oct 1814	12th US Inf	Capt James Paxton
Bowman John	28 Oct 1814	12th US Inf	Capt James Paxton
Grammer Robert	21 Jun 1813	12th US Inf	Capt James Paxton
Helms Abraham	14 May 1814	12th US Inf	Capt James Paxton
James Jacob	13 Mar 1814	12th US Inf	Capt James Paxton
Manly Bay	29 Oct 1814	12th US Inf	Capt Post
Manly Littlebury	29 Oct 1814	12th US Inf	Capt Andrew Madison
Pully Robert	22 Nov 1814	12th US Inf	Capt Willis
Shaw Nathaniel	24 Oct 1814	12th US Inf	Unknown
Turk Benjamin	25 Dec 1813	13th US Inf	Capt M M Dox
Van Sickler Gilbert	25 Dec 1813	13th US Inf	Capt Myndert Dox
Ausant William	10 Jan 1815	14th US Inf	Capt Earle
Austin Benjamin	18 Jan 1815	14th US Inf	Capt Gilder
Brazur Thomas	10 Jun 1815	14th US Inf	Capt Marshall
Browson Oron	Nov 1814	14th US Inf	CaptJ Stanard
Hagan Isaac	8 Feb 1815	14th US Inf	Capt Earle
Lecompte William	10 Jan 1815	14th US Inf	Capt Richard Arell
Peach Issac	8 Feb 1815	14th US Inf	Unknown

Pryor John	22 Nov 1814	14th US Inf	Capt Gilder
Rictor Christian	Nov 1814	14th US Inf	Capt Gilder
Flemming James	3 Nov 1814	15th US Inf	Capt Hayden
Reed Joseph	9 Jan 1815	15th US Inf	Capt H H VanDalsem
Sumners John	24 Oct 1814	15th US Inf	Capt Barton
Talmadge Robert	9 Jun 1815	15th US Inf	Capt H H VanDalsen
Terhune Albert	15 Jan 1815	15th US Inf	Capt Hayden
Barrett Levi	14 Dec 1814	16th US Inf	Capt J Baldy
Bell Joseph	Feb 1814	16th US Inf	Capt J T McElroy
Matson John	10 Dec 1814	16th US Inf	Capts Atkins & Mc Laughlin
Williams john	13 Dec 1814	16th US Inf	Capt Atkins
Alexander Benjamin	10 Jan 1815	17th US Inf	Capts Jacksons & Saunders
Ball Adam	7 Jan 1815	17th US Inf	Capt David Holt
Ballard Linsey	1 Mar 1815	17th US Inf	Capt Trimball
Benjamin Alexander	10 Jan 1815	17th US Inf	Capt H H Hickman
Berkley John	30 May 1815	17th US Inf	Capts Adair & CaptHolt
Brown John	11 Jan 1815	17th US Inf	Capt Adair
Butler Thomas	21 Dec 1814	17th US Inf	Capt David Holt
Campbell Samuel	Feb 1814	17th US Inf	Capt H H Hickman
Canell Castle	21 Dec 1814	17th US Inf	Lt Hackley
Capil Samuel	9 Dec 1814	17th US Inf	Lt Hackley
Capitte Paul	9 Dec 1814	17th US Inf	Capt David Holt
Crawford Michael	7 Dec 1814	17th US Inf	Capt David Holt
Dan Peter	14 Jan 1815	17th US Inf	Capt Hackley
Darnell Samuel S	1 Feb 1815	17th US Inf	Capt Benjamin Sanders
Den Pilee	15 Feb 1815	17th US Inf	Capt H H Hickman
Derr Peter	Feb 1815	17th US Inf	Capt H H Hickman
Duncan George	10 Aug 1814	17th US Inf	Capt John T Chunn
Flemming David	14 Jan 1815	17th US Inf	Capt H H Hickman
Forlorn Meredith	21 Jan 1815	17th US Inf	Capt Mordecai Myers
Fryman Henry	Unknown	17th US Inf	Capt H H Hickman
Hardisty George	11 Feb 1815	17th US Inf	Unknown
Hopkins Isom	15 July 1815	17th US Inf	Capt Scott
Johns John	25 Mar 1815	17th US Inf	Capt David Holt
Kennel Casper	21 Dec 1814	17th US Inf	Capt H H Hickman
Layor John	28 Feb 1815	17th US Inf	Capt William J Adair
Lemmon Benjamin	27 Dec 1814	17th US Inf	Capt David Holt
Marsh William	29 Jan 1815	17th US Inf	Capt H Hickman
Mays John	31 Dec 1814	17th US Inf	Capt John T Chunn
Mc Amy Samuel	15 Jan 1815	17th US Inf	Capt John T Chunn
Mc Combs John	17 June 1814	17th US Inf	Capt Trimbal
Mc Donald Stephen	10 Jan 1815	17th US Inf	Capt David Holt
Moules James	15 Jan 1815	17th US Inf	Unknown
Nation Frederick	21 March 1815	17th US Inf	Capt Henry Crittenden
Norris Beverly	25 Dec 1814	17th US Inf	Capt Benjamin Sanders
Palmer Richard	8 Dec 1814	17th US Inf	Capt H H Hickman
Parmer Richard	8 Dec 1814	17th US Inf	Unknown

Patterson Alexander Lt	Unknown	17th US Inf	Unknown
Robbins Hiram	13 Jan 1815	17th US Inf	Unknown
Roberts Hiram	Jan 1815	17th US Inf	Capt H H Hickman
Sayor John	1 Mar 1815	17th US Inf	Capt Adair
Sidebottom Wilson	18 Mar 1815	17th US Inf	Capt Adair
Smith Joel	24 Dec 1814	17th US Inf	Lt Holt
Sprague Ephraim	25 Jan 1815	17th US Inf	Capt Hickman
Stewart Alex	7 Dec 1814	17th US Inf	Capt B W Sanders
Bear Adam	1 Dec 1814	19th US Inf	Capt Carey A Trimble
Bledsoe Jacob	15 Oct 1814	19th US Inf	Capt William McDonald
Denman Samuel	30 Dec 1814	19th US Inf	Capt Carey A Trimble
Duncan Samuel	6 Jan 1814	19th US Inf	Capt William McDonald
Flemming Patrick	5 Nov 1814	19th US Inf	Capt John T Chunn
Gluffan Thomas	22 Oct 1815	19th US Inf	Capt William McDonald
Griffin Thomas	22 Oct 1814	19th US Inf	Capt William McDonald
Mc Clintock John	13 Sep 1814	19th US Inf	Lt Riddles
Moles James	26 Jan 1815	19th US Inf	Capt H H Hickman
Stansbury Christian	Jan 1815	19th US Inf	Capt Carey A Trimble
Topping Zephar	27 Sep 1814	19th US Inf	Capts McDonald & Trimble
Davis Willis	Sept/Oct 1813	20th US Inf	Capt J Macrae
Franklin Joseph	30 Dec 1813	20th US Inf	Capt John Stanard
Mills William	1 Apr 1813	20th US Inf	Lt T M Randolph
Ames Daniel	1 Jan 1814	21st US Inf	Capt John T Chunn
Baker Amos	4 Oct 1814	21st US Inf	Capt Sullivan Burbank
Bidman/Bidmore Peter	30 Oct 1814	21st US Inf	Cap tWillis Foulk
Grouard James	5 May 1815	21st US Inf	Unknown
Hibbert Jeremiah	15 Feb 1815	21st US Inf	Capt James Pratt
Hutchings William	25 Feb 1815	21st US Inf	Capt Lemuel Bradford
Ireland Joel	Unknown	21st US Inf	Capt Lemuel Bradford
Lincoln Peris	26 Sept 1814	21st US Inf	Capt Sullivan Burbank
Littlefield George Sergeant	1 Feb 1815	21st US Inf	Capt Benjamin Ropes
Millhallen Daniel	Unknown	21st US Inf	Capt Henderson
Mudge Charles	19 Aug 1814	21st US Inf	Capt Sullivan Burbank
Packard/Packer George	Oct 1814	21st US Inf	Capt Sullivan Burbank
Pattee Ezekeil	4 Nov 1814	21st US Inf	Capt Morrill Marston
Peckan George	27 Nov 1814	21st US Inf	Capt Benjamin Ropes
Pettingall William	16 July 1816	21st US Inf	Capt Cass
Petty Ezekial	4 Nov 1814	21st US Inf	Capt Morrill Marston
Philbrook Jonathan	18 Dec 1814	21st US Inf	Capt Joseph Treat
Priest Bartholemew	27 Aug 1814	21st US Inf	Capt Bowman
Smith Charles Sergeant	11 Oct 1814	21st US Inf	Capt Morrill Marston
Smith John	10 Oct 1814	21st US Inf	Capt Joseph Treat
Smith John	28 Oct 1814	21st US Inf	Capt William Bowman
Spearin James	Unknown	21st US Inf	Capt Lemuel Bradford
Stewart John	24 Apr 1815	21st US Inf	Unknown
Stewart Lemuel	12 Dec 1814	21st US Inf	Capt Joseph Treat
Tenney Samuel	18 Jan 1815	21st US Inf	Capt Morrill Marston

Thomson Curtis	11 Nov 1813	21st US Inf	Capt Joseph Treat
Wheddon Samuel	29 Jan 1815	21st US Inf	Capt Ira Drew
Whidden Timothy	25 Feb 1815	21st US Inf	Capt Lemuel Bradford
Witham Levi	1 Dec 1814	21st US Inf	Capt Marston
Clark Thomas	17 Oct 1814	22nd U S Inf	CaptJ oseph Henderson
Armstrong John Lieutenant	28 Aug 1814	22nd US Inf	Capt Willis Foulk
Bell James	30 Oct 1814	22nd US Inf	Unknown
Billman Peter	30 Oct 1814	22nd US Inf	Capt James Henderson
Bowhill John	Unknown	22nd US Inf	Capt William Morrow
Brady Hugh Colonel	Unknown	22nd US Inf	Unknown
Byron Edward	25 Aug 1814	22nd US Inf	Capt Jos Henderson
Caldwell Andrew	30 Oct 1814	22nd US Inf	Col Hugh Brady
Cannon Martin	10 Sept 1814	22nd US Inf	Colonel Hugh Brady
Carman Martin	10 Sept 1814	22nd US Inf	Capt Morris or Henderson
Cochran Willam	1 Oct 1814	22nd US Inf	Unknown
Coughman William	Oct 1814	22nd US Inf	Capt Thomas Lawrence
Crouse John Corporal	20 Sept 1814	22nd US Inf	Capt Anderson
Cummings James	11 Sept 1814	22nd US Inf	Lt John R Guy & Willis Foulk
Gladin William	20 Feb 1815	22nd US Inf	Capt Willis Foulk
Griffy John	10 Dec 1814	22nd US Inf	Lt Richards
Guthrie James	5 Dec 1814	22nd US Inf	Capt Sampson L King
Hammon William	11 Sept 1814	22nd US Inf	Capt Sampson L King
Harble Anthony	1 Oct 1814	22nd US Inf	Capt Willis Foulk
Harman George	11/17 Sept 1814	22nd US Inf	Capt Joseph Henderson
Harmon William	17 Sept 1814	22nd US Inf	Capt Joseph Henderson
Hawkesbury Adam	2 Sept 1814	22nd US Inf	Capts Henderson & Cormack
Hickman William Sergeant	1814	22nd US Inf	Capt John Foster
Hock Grunby Sergeant	14 Oct 1814	22nd US Inf	Capt Willis Foulk
Hook Greenbury Sergeant	14 Oct 1814	22nd US Inf	Capt Willis Foulk
Hoover Jacob Corporal	26 Oct 1814	22nd US Inf	Capt Joseph Henderson
Izard Jonathan	Oct 1814	22nd US Inf	CaptJohn R Guy
Lowre John	19 Sep 1814	22nd US Inf	Lt John Brady
Martial/Marshall George	17 Oct 1814	22nd US Inf	Unknown
McDay Edward	27 Oct 1814	22nd US Inf	Capt Joseph Henderson
Mcllhenny William	11 Jan 1815	22nd US Inf	Capt Willis Foulk
Noggle Peter	19 Aug 1814	22nd US Inf	Capt Sampson L King
Prier John	23 Oct 1814	22nd US Inf	Capt Sampson L King
Ramby/Ramsey Nathan	2 Nov 1814	22nd US Inf	Capts Morrow & Henderson
Reddick George	19 Feb 1815	22nd US Inf	Capt William Morrow
Rediur/Redier Jacob	31 Oct 1814	22nd US Inf	Capt Joseph Henderson
Ricker John	5 Nov 1814	22nd US Inf	Capt Joseph Henderson
Rothwell Jacob	Unknown	22nd US Inf	Capt D Espy
Savage Samuel	30 Oct 1814	22nd US Inf	Capt William Morrow
Smith William	16 Sep 1814	22nd US Inf	Capt Joseph Henderson
Stout David	18 Sept 1814	22nd US Inf	Capts Morris & Henderson & King
White George	13 Sep 1814	22nd US Inf	Capt Thomas Lawrence
Wright George	15 Sep 1814	22nd US Inf	Capt Sampson L King

Moore Alexander Corporal	23 Aug 1814	22nd US Inf	Capt Willis Foulk
Moore Charles	Dec 1814	22nd US Inf	Capt John Pentland
Guy John	7 Sept 1814	23nd US Inf	Capt Ingersoll
Bennett Paul	21 Sept 1814	23rd US Inf	Unknown
Brown Walter	4 Jan 1815	23rd US Inf	Capt Richard Goodell
Burrell William	31 Oct 1814	23rd US Inf	Capt F Brown
Cooper Lewis	Unknown	23rd US Inf	Capt Azariah W Odell
Gay John	7 Sept 1814	23rd US Inf	Lt Justus Ingersoll
Holdin Rufus	17 Sept 1814	23rd US Inf	Capt Richard Goodell
Johnson Jonathan	19 Sept 1814	23rd US Inf	Lt Justus Ingersoll
Lawrence Benjamin Corporal	29 Dec 1814	23rd US Inf	Lt Castine
Mc Donald David	13 Sep 1814	23rd US Inf	Capt Justus Ingersoll
Miller Solomon	4 Oct 1814	23rd US Inf	Capt Whiting
O' Riley George	25 Sep 1814	23rd US Inf	Capt Richard Goodell
O' Riley John	6 Sep 1814	23rd US Inf	Capt Richard Goodell
Quimby James	27 Oct 1814	23rd US Inf	Capt Abraham Hull
Ripley	5 Nov 1814	23rd US Inf	Capt Simeon D Wattles
Salisbury Simon	11 Sept 1814	23rd US Inf	Capt Samuel Tappan
Thayer Nathaniel	23 Aug 1814	23rd US Inf	Capt Justus Ingersoll
Weeks Samuel	27 Sep 1814	23rd US Inf	Capt Samuel Tappan
Wiley Henry	13 Nov 1814	23rd US Inf	Lt Samuel Spafford
Keeler John	13 Jan 1815	24th US Inf	Capt John T Chunn
Brace Stephen Jr	7 Jan 1815	25th US Inf	Lt Thomas Seymour
Braman Daniel	31 Oct 1814	25th US Inf	Capt Daniel Ketchum
Champlin George	30 Aug 1814	25th US Inf	Capt Daniel Ketchum
Clark Shaderack	1 Nov 1814	25th US Inf	Lt Thomas Seymour
Cook Aaron	23 Dec 1814	25th US Inf	Capt Daniel Ketchum
Corning David	31 Aug 1814	25th US Inf	Capt Daniel Ketchum
Cottar Joseph	15 Feb 1815	25th US Inf	Lt Thomas Seymour
Hard Hiram/Niram	28 Aug 1814	25th US Inf	Unknown
Haskins Levi	7 Oct 1814	25th US Inf	Capt Daniel Ketchum
Hastings Joseph	19 Sept 1814	25th US Inf	Capt George Howard
Howard Jonathan	7 Nov 1814	25th US Inf	Lt Thomas Seymour
Jennings Noah	11 May 1814	25th US Inf	Capt John B Murdock
Knowlton John	19 Sept 1814	25th US Inf	Capt Seymour
Liswill Thomas	16 Nov 1814	25th US Inf	Capt Thomas Seymour
Lomis Samuel	7 Oct 1814	25th US Inf	Capt Daniel Ketchum
Mansfield Samuel	27 Sep 1814	25th US Inf	Capt Benjamin Watson
Marcellus Cornelius	19 Aug 1814	25th US Inf	Lt Edward White
Miller Asahel Sergeant	5 Oct 1814	25th US Inf	Capt Thomas Seymour
Nelson Silas	26 Oct 1814	25th US Inf	Capt Daniel Ketchum
Potter James Sergeant	22 Aug 1814	25th US Inf	Capt George Howard
Rusco Gamlien	6 Aug 1814	25th US Inf	Lt Thomas Seymour
Schofield Joseph	19 Sept 1814	25th US Inf	Capt Seymour
Scott John	23 Aug 1814	25th US Inf	Capt Daniel Ketchum
Smith Reuben	23 Sept 1814	25th US Inf	Capt Daniel Ketchum
Story Nathan Corporal	10 Nov 1814	25th US Inf	Capt Daniel Ketchum

Stoughton Russel Jr	23 Sept 1814	25th US Inf	Capt Daniel Ketchum
Taylor Elisha	25 Oct 1814	25th US Inf	Capt Daniel Ketchum
Taylor Peter	1 Nov 1814	25th US Inf	Lt Thomas Seymour
Tuttle Charles	6 Aug 1814	25th US Inf	Capt Daniel Ketchum
Waldron Nicholas	25 Feb 1815	25th US Inf	Capt Thomas Seymour
Weed Scridder/Scudder	29 Sep 1814	25th US Inf	Lt Thomas Seymour
Winchel Shaylor	27 Sep 1814	25th US Inf	Capt Daniel Ketchum
Wright Judah	1 Nov 1814	25th US Inf	Capt Daniel Ketchum
Hayette Elisha	15 Oct 1814	27th US Inf	Capt Charles L Cass
Robinson Zadock	22 Dec 1814	Engineers	Capt J W Goodin
Currier Joseph	11 Sept 1814	Sappers	Capt David B Douglas
Matchel/Matchett Robert	26 Dec 1814	1st US Arty	Unknown
Banister Luther	Apr 1815	US Arty	Capt G W Melvin
Bostwick Richmond - Sergeant	2 Sept 1814	US Arty	Capt Alex Williams
Bower Andrew	15 Jan 1815	US Arty	Capt Crary
Bowers Andrew	15 Jan 1815	US Arty	Capt H K Craig
Bowling Simon	17 Feb 1815	US Arty	Capt A C W Fanning
Brickman Joseph	6 Oct 1814	US Arty	Capt Thomas Biddle
Broadwater David	31 Oct 1814	US Arty	Capt Alex J Williams
Calvin William	31 Oct 1814	US Arty	Capt Thomas Biddle
Carter Jacob	14 Sept 1814	US Arty	Capt Thomas Biddle
Colvin William	31 Oct 1814	US Arty	Capt Thomas Biddle
Custer Jacob	Sept 1814	US Arty	Capt Thomas Biddle
Giberson William	Sept 1814	US Arty	Capt Thomas Biddle
Gibson Nathanial	9 Sept 1814	US Arty	Capt John Ritchie
Gibson Thomas Corporal	11 Sept 1814	US Arty	Capt Richie
Girton Jonathan	31 Dec 1814	US Arty	Capt Samuel B Archer
Giverson William	7 Oct 1814	US Arty	Lt Douglas
Goodwin Peter	11 Sept 1814	US Arty	Capt Thomas Biddle
Hickman James	2/3 Jan 1815	US Arty	Capt A J Williams
Jamison Rich	18 Mar 1815	US Arty	Capt Alexander Williams
Morrell Joseph	24 Oct 1814	US Arty	Col Scott & Williams
Paselo Henry	10 Sep 1814	US Arty	Capt Thomas Biddle
Pierce John	23 Aug 1814	US Arty	Capt A J Williams
Pike Artemas	26 Oct 1814	US Arty	Capt Thomas Biddle
Randon Paul	6 Apr 1815	US Arty	Capt Alex J Williams
Stiles Eli	1 Sept 1814	US Arty	Capt A J Williams
Sutter Tobat/Tobal	1 Oct 1814	US Arty	Capt Thomas Biddle
May John F	7 Dec 1814	US Arty	Capt Thomas Ketchum
Morrill Moses	11 Aug 1814	US Atry	Capt Thomas Biddle
Seeley Jeremiah	14 Sept 1814	US Atry	Capt John W Goodins
Skinner Joshua	9 Feb 1815	US Atry	Capt J W Goodin
Vosburgh Russell	16 Mar 1814	US Atry	Capt A W Thorton
Buckman Francis	2 Mar 1814	US Lt Arty	Capt Thornton
Burkans Simon	24 Mar 1814	US Lt Arty	Capt Branch
Foster Luther C	26 Dec1813	US Lt Arty	Capts Branch & Stribling
Baker D	1 Oct 1814	US Lt Dragoons	Capt Haynes

Baldwin Nathaniel	3 Mar 1815	US Lt Dragoons	Capt Haig
Blay Francis	27 Jan 1815	US Lt Dragoons	Capt Samuel D Harris
Lint Lewis	13 Jan 1815	US Lt Dragoons	Capt Large
Holder Daniel	Unknown	US Rifles	Capt William Smyth
Blodgett Clark	13 Sept 1814	NY milita	Capt Moody R Freeman
Hewes William	4 Aug 1814	NY milita	Capts Freeman & Valentine
Paul John	25 Sep 1814	NY milita	Capt Nathaniel F Knapp
Staples Joshua	25 Sept 1814	NY Militia	Capt Joshua Harkin
Graves Lebins	25 Sept 1814	NY or PA Vol	Unknown
Norris Joseph	30 Aug 1814	NY or PA Vol	Capt Brooks
Shells Henry	9 Oct 1814	NY or PA Vol	Unknown
Grey Duncan	25 Sept 1814	NY Vol	Capt Henry Parker
Green Nathanial	9 Sept 1814	NY Vol Cavalry	Capt Haig
Bocher Henry	23 Aug 1814	PA Vol	Capt Samuel Gorden
Brown Alex	16 Sept 1814	PA Vol	Capt James Piper
Cooper Thomas	14 Aug 1814	PA Vol	Capt Andrew Mitchell
Kelley Morgan	30 Aug 1814	PA Vol	Capt Andrew Mitchell
Mawheter William	16 Sep 1814	PA Vol	Capt John Roberts
Woodburn Robert	4 Sep 1814	PA Vol	Capt James Piper
Myers George	12 Sep 1814	PA Vol	Cap tSamuel White
Abbott Daniel L	13 Mar 1814	Unknown	Unknown
Appleby Nathan	16 Oct 1814	Unknown	Capt Foster
Barney John	4 Oct 1814	Unknown	Unknown
Boynton Abel Colonel	2 Sept 1814	Unknown	Unknown
Braden George	21 May 1814	Unknown	Capt Moore
Burrans Semor	24 Mar 1814	Unknown	Capt McIntire
Collin James	30 May 1814	Unknown	Capt Goodridge
Curthburgh	16 Feb 1814	Unknown	Capt McIntire
Elwell Robert	18 Apr 1814	Unknown	Capt Barker
Frick David	4 Oct 1814	Unknown	Capt Cobley
Gale John	7 May 1814	Unknown	Capt Larabee
Griffin Theophiles	14 Sept 1814	Unknown	Capt Hawkins
Hull Simon	July 1814	Unknown	Capts Weeks & Hull
Johnson David	7 Apr 1814	Unknown	Capt VanDolson
Lewis John	2 Mar 1814	Unknown	Capt Thorington
Lewis Lewis	24 May 1814	Unknown	Capt Mutchenburgh
Mann John	22 Mar 1814	Unknown	Capt Barker
Martin Thomas	29 Sep 1814	Unknown	Capt Spencer
Mc Nally John	25 May 1814	Unknown	Capt Muhlenberg
Morton Obed Sergeant	9 Dec 1814	Unknown	Capt Launders
Nichols James	15 Oct 1814	Unknown	Unknown
Perkans James	6 Oct 1814	Unknown	Capt Eastman
Pummons John	24 Oct 1814	Unknown	Capt Bartons
Smith Elias	30 Mar 1814	Unknown	Capt Fuller
Smith Lyman Corporal	23 Sept 1814	Unknown	Capt Douglas
Stickney Abijah	30 Sept 1814	Unknown	Capt Mumford
Swany Edward	23 May 1814	Unknown	Capt Branches

Thompson Thomas	8 Apr 1814	Unknown	Capt McCoon
Thrahal Jacob	1 Mar 1814	Unknown	Unknown
Tibbitts John	5 Sept 1814	Unknown	Unknown
Tibbs Richard	Unknown	Unknown	Capt Moore
Toby Thomas	27 Feb 1814	Unknown	Capt Ball
Twiss Samuel	30 Sept 1814	Unknown	Capt Stridin
Wells Stephen	5 Oct 1814	Unknown	Capt White
Wheeton William	19 Oct 1814	Unknown	Capt Hawkins
White John	13 Mar 1814	Unknown	Capt Thorington
Young Nathaniel	16 Nov 1814	Unknown	Capt Kinney

Also, many British Soldiers Known Only Too God.

8
GARRISON BURIAL GROUNDTIME LINE

Based on War Department Records, this cemetery along Aero Drive became a mass grave for American Soldiers beginning with William Hewes on 4 August 1814 and ending with William Pettingall on 16 July 1815. It is most likely that many more American soldiers were buried in the mass grave, but the records did not survive. They also buried an unknown number of British soldiers captured during the Niagara Campaign of 1814 on this site.

The Holland Land Company's first land purchase for the small military cemetery in Lot 7, Township 11, Range 7 was originally purchased as part of 202 acres of land on 8 October 1808, by Frank Ludlow, who owned it until 11 January 1816 (Livsey, 1991).

Then, on 11 January 1816, Wilhem Willink and colleagues (Holland Land Co.), sold property containing the cemetery to Benjamin Barton. (Liber 3 page 133)

17 October 1821, Barton sold property containing the cemetery to Thomas Coe. (Liber 7 page 207)

4 July 1851 William Coe, sold property containing the cemetery to John Haskell. (Liber 172, page 459)

. . .

6 August 1851 John Haskell, sold the cemetery property only (parcel A) to John Bliss and John R. Evans (Liber 123 page 19) as, "The ground should not be disturbed or encroached upon."

12 July 1898, the estates of John Bliss and John R. Evans deeded the cemetery (Parcel A) to the Buffalo Historical Society. (Liber 833 page 283)

9 November 1898, the Buffalo Historical Society placed a cannon on the burial site. The cannon was a Parrott Gun taken from the north end of the circle at The Front and was suitably inscribed to the memory of the dead soldiers. They installed an arched gateway with the letters **Soldiers of the War of 1812** cut from galvanized iron.

By 1918, the old cemetery had fallen into disrepair. The Buffalo Historical Society responded by making improvements again to the cemetery.

In 1932 after much research, the Buffalo Historical Society located the hospital register which was in the possession of the federal government and retrieved the names of over 200 American soldiers who died in the General Military Hospital. It began on 1 August 1814 and the remains were buried at the cemetery. They did not find the names of the soldiers buried before that date until 2011 (see entry for 2011).

30 May 1933, a Memorial Day service was held to rededicate the cemetery in honor of the soldiers now identified and buried in the cemetery. The Buffalo Historical Society made improvements to the burial site, removing the arch and adding boulders with plaques inscribed as follows:

(1) This property is owned and maintained by the Buffalo Historical Society

(2) This tablet is erected by the Buffalo Historical Society in honor of the American and British soldiers of the War of 1812, who died in the U.S. Military Hospital at Williamsville, New York and maintained by the Buffalo Historical Society

In 1936, the Frontier Post #71, Canadian Legion of the British Empire Service League in Fort Erie Ontario, presented a cannon captured from the Americans in the War of 1812.

. . .

30 May 1939, a ceremony was held by the Buffalo Historical Society to dedicate the new cannon. A carriage had been built for the cannon which was placed in the middle of the cemetery along with two flag poles erected for American and British flags.

18 June 1939, the Town of Cheektowaga sponsored a dedication at the cemetery on the occasion of the town's centennial celebration.

25 August 1942, the original cannon (acquired in 1898), was presented to the Bethlehem Steel Corporation to be used as scrap metal for the war.

During the 1940s and 50s, Frank Roman of Ellicott Creek Road cared for the cemetery on behalf of the Buffalo Historical Society. In addition, the George F. Lamm Post remembered the veterans by placing flags at the graves on occasion.

17 June 1962, Cheektowaga Post #2429 VFW along with the Buffalo Historical Society rededicated the cemetery for the 150th anniversary of the War of 1812. American and Canadian units took part, and they added a monument with the inscription:

War of 1812 - 150 Years - 1962 Dedicated June 17, 1962, to the valor of the American and British Soldiers, and the peace that has long endured

During the 1960s, Cheektowaga Post #2429 VFW took responsibility for the upkeep of the cemetery.

5 October 1975, Cheektowaga celebrated the U.S. Bicentennial with representatives from both the United States and Canada taking part in ceremonies at the burial site.

In October, 1984 Cheektowaga took the first steps in acquiring the cemetery with a firm hired to do the search and survey.

. . .

On 28 January, 1985 the Cheektowaga town Board approved a resolution to acquire the cemetery property. The town agrees to "maintain the War of 1812 Cemetery and to forever dedicate such premises as a cemetery and resting for the soldiers who fought in the War of 1812"

20 March, 1985 a quit-claim deed is signed conveying the cemetery (parcel A) from the Buffalo and Erie County Historical Society to the Town of Cheektowaga. (Liber 9444 page 450)

On 21 March, 1985 the Cheektowaga Town Board approves a resolution transferring ownership of the cannon to the town. "The gift was made subject to the condition that the cannon be perpetually maintained and be forever dedicated as a memorial to the British and American soldiers who fought during the War of 1812".

23 May, 1985 a quit claim deed is signed conveying additional cemetery property (parcel B) from Pricilla Brown to the Town of Cheektowaga. (Liber 9450 page 680)

On 9 June, 1985 a dedication is held to commemorate the conveyance of the cemetery to the Town of Cheektowaga. The re dedication included representatives from both the United States and Canada. A letter was received from Queen Elizabeth II commending the town.

In 2002 The cemetery is listed on the National Register of Historic Places.

2011 Jack Bilow publishes "A War of 1812 Death Register" after searching the National Archives for six months. The register adds the names of the soldiers buried at Williamsville prior to 1814.

9

THE LOST CANTONMENT OF WILLIAMSVILLE

The story that follows concerns a chapter of the War of 1812 that has been almost entirely forgotten an encampment built in haste, contested in court, and then swallowed by time. Most people in Western New York know of the Williamsville War of 1812 Barracks and Hospital. Far fewer, however, are aware of the principal encampment: Cantonment Williams Ville.

This cantonment survives today only in scattered references buried in manuscripts, letters, and land records. To reconstruct its history, we must imagine the winter of 1812: riflemen ordered to build huts on land that did not belong to them; officers acting under pressure; and a village landscape transformed overnight.

When Jonas Williams returned from Albany, he found his finest timber cut and an Army cantonment erected on his property. He was furious—and powerless. He sued multiple officers for thousands of dollars. Yet amid this legal turmoil, a sprawling complex of log huts rose along the creek, remained throughout the war, and then vanished almost as quickly as it had appeared.

For two centuries, the exact footprint of this encampment has been lost. But the most compelling part of the story is not the huts themselves. It is what may have been left behind. As I traced the Hopkins papers, the Coles and Schuyler letters, and the shifting land ownership, a pattern emerged—one that pointed not only to the cantonment's location but also to something far more solemn: a burial ground.

This burial place was mentioned only in passing, never mapped, never marked, and never connected to the Rifle Cantonment until now.

We still cannot say with certainty where that burial ground lies. The evidence is strong but incomplete. The clues align, but they do not yet lock into place. What we have is a mystery—one that has waited more than two hundred years for someone to notice it. In the pages that follow, I will walk through the documents, the land, and the logic that bring us closer than anyone has ever been to understanding where the riflemen camped—and where some of them may have been laid to rest. This is the story of a cantonment nearly erased, and a burial ground that refuses to disappear.

The Overlooked Cantonment Within the Village

When we look at the familiar historical marker at Main and Evans, most people do not realize that it refers to the Artillery Cantonment, later converted into hospital huts and eventually replaced by a General Hospital. What even fewer people know is that a second, much larger cantonment once stood within the village itself.

How do we know this, especially when it appears in no local summaries or signage?

The answer lies in the earliest stage of site selection for the War of 1812 encampments. When General Smyth instructed Colonels Schuyler and Coles to identify a suitable location, they chose Williamsville for two practical reasons: its reliable spring water and its working sawmill both essential for sustaining a large military force.

But here is the twist: the land they selected belonged to Jonas Williams, who at that moment was serving in Albany as a member of the state legislature. The officers chose his property without his knowledge or consent. Several early documents consistently reference a main body of troops stationed within the village itself, restoring a forgotten chapter of Williamsville's wartime landscape.

Eyewitness Accounts: Hopkins and the Early Encampment

One of the earliest and most valuable descriptions comes from the manuscript of Timothy A. Hopkins, housed at the Buffalo History Museum. Hopkins recalled that shortly after the declaration of war, approximately six thousand troops were ordered to Williamsville for winter quarters, many remaining until the end of hostilities.

He described the main body of soldiers stationed on the north side of Main Street, where long rows of log barracks stretched from the village spring—later the site of the brewery—eastward toward Ellicott Creek. Hopkins also recounted the moment Jonas Williams returned to find his land transformed into a military installation, his timber gone, and his property commandeered without consent.

This account aligns with the lawsuits that soon followed.

From the Timothy A. Hopkins Papers Buffalo Museum of History

Colonel Coles: A Prisoner Over a Cantonment

A remarkable letter written by Colonel I. A. Coles on January 1, 1814, reveals the personal consequences of the cantonment's construction. Writing from confinement in New York, Coles explained that he had been ordered by General Smyth, through Colonel Parker, to select and clear land near Williamsville for winter quarters. He carried out the order—only to be sued by Jonas Williams for seven thousand dollars, an enormous sum at the time.

Coles wrote that he was now held as a "close prisoner" while awaiting the government's decision on his case. His postscript identifies Williams by name and notes that similar legal actions had been brought against other officers, including Major Huyck and Colonel Schuyler.

This letter confirms not only the cantonment's location but also the intensity of the conflict it created.

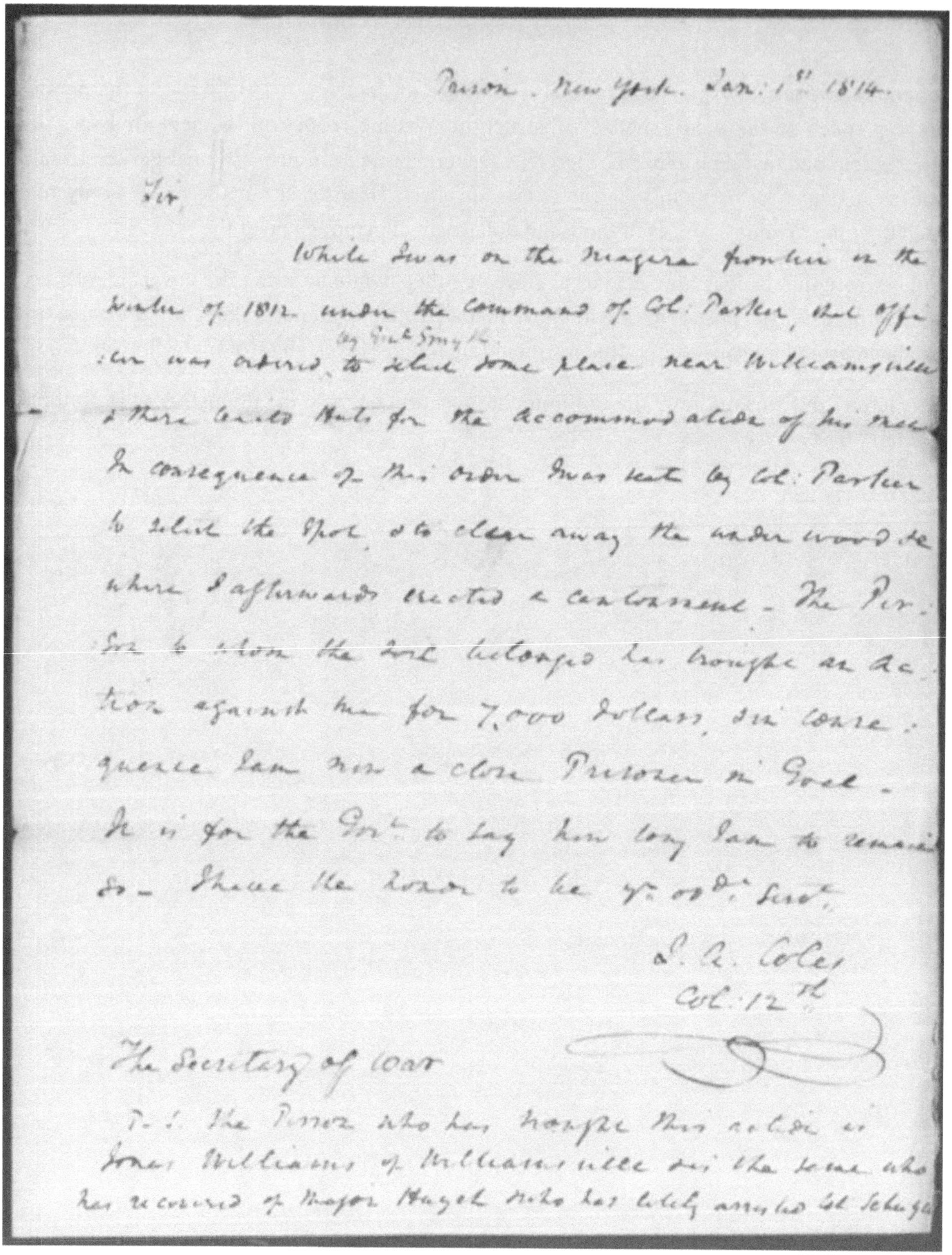

Prison. New York. Jan: 1st 1814.

Sir,

While I was on the Niagara frontier in the winter of 1812 under the command of Col. Parker, that officer was ordered by Genl Smyth to select some place near Williamsville & there build Huts for the accommodation of his men. In consequence of this order I was sent by Col: Parker to select the Spot, & to clear away the under wood &c where I afterwards erected a Cantonment. The Person to whom the land belonged has brought an Action against me for 7,000 dollars, & in consequence I am now a close Prisoner in Goal. It is for the Govt. to say how long I am to remain so. I have the honor to be Yr. obdt. Servt.

I. A. Coles
Col: 12th

The Secretary of War

P.S. The Person who has brought this action is Jonas Williams of Williamsville and the same who has received of Major Hugh who has lately arrested Col. Schuyler

Col. I. A. Coles Letter to the Secretary of War January 1, 1814

Prison, New York, Jan. 1st, 1814

While I was on the Niagara frontier in the winter of 1812 under the command of Col. Parker that officer was ordered by General Smyth or was ordered to select some place near Williamsville & there build Huts for the accommodation of his men In consequence of this order was sent by Col. Parker to select the spot & to clear away the underwood & where I afterwards erected a cantonment. The Person to whom the soil belonged has brought an action against me for 7,000 dollars & in consequence I am now a close Prisoner in God It is for the Govt to say how long I am to remain so. - I have the honor to be your servant.

I. A. Coles Col. 12th

The Secretary of War

P.S. The Person who has brought this action is Jonas Williams of Williamsville he's the same who has recorded of Major Huyck who has lately arrested Col. Schuyler

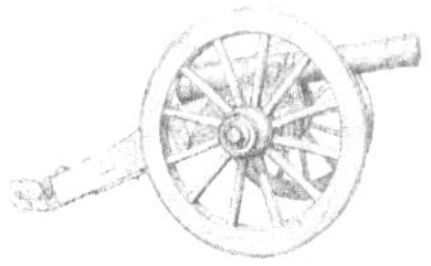

Colonel Schuyler's Testimony: The Eleven Mile Creek Cantonment

Another critical document is Colonel P. P. Schuyler's letter to the Secretary of War dated November 19, 1814. Schuyler reported that he, too, had been sued for trespassing on Williams's land in 1812 while constructing the cantonment at Eleven Mile Creek.

He described the circumstances in detail: after General Smyth abandoned plans to invade Canada, officers were ordered to "hut for the winter." With the season far advanced and no suitable materials near their existing encampment, the Eleven Mile Creek site became, in Schuyler's words, "a dernier resort."

Schuyler and Coles selected the ground, informed Williams's agent of their intentions, and assured him that the government would compensate the owner. Major Huyck was ordered to clear the land, and the regiments soon followed. By 1814, the site had been continuously occupied by U.S. troops, and the damages to Williams's property were described as "very considerable."

Schuyler urged the War Department to intervene before the case reached the Supreme Court.

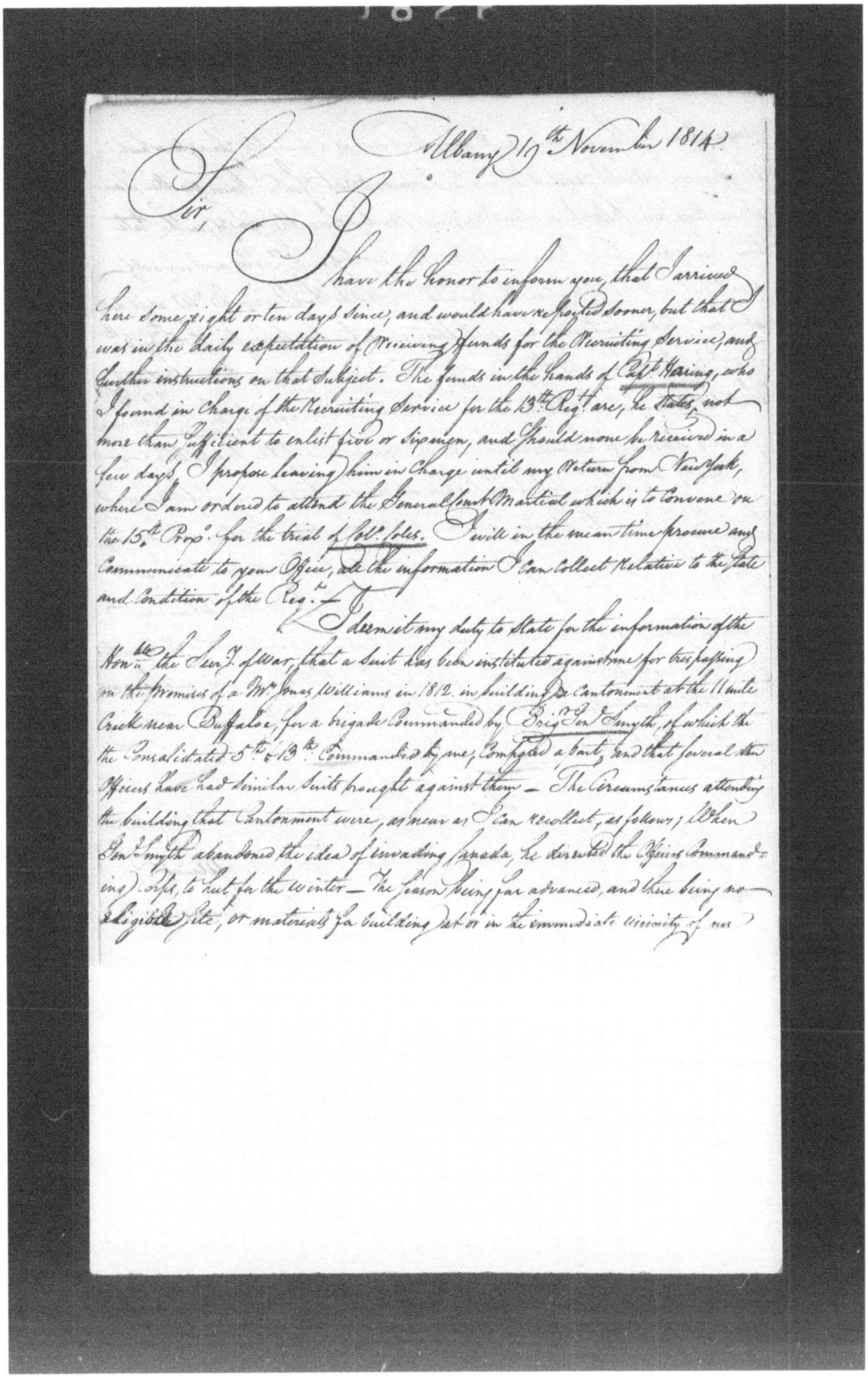

Albany 19th November 1814.

Sir,

I have the honor to inform you, that I arrived here some eight or ten days since, and would have reported sooner, but that I was in the daily expectation of receiving funds for the Recruiting Service, and further instructions on that Subject. The funds in the hands of Capt. Haring, who I found in Charge of the Recruiting Service for the 13th Regt. are, he states, not more than sufficient to enlist five or six men, and should none be received in a few days, I propose leaving him in Charge until my Return from New York, where I am ordered to attend the General Court Martial which is to convene on the 15th Prox. for the trial of Col. Coles. I will in the mean time procure and communicate to your Office, all the information I can collect relative to the State and Condition of the Regt.

I deem it my duty to State for the information of the Honble the Secy. of War, that a Suit has been instituted against me for trespassing on the premises of a Mr. Jonas Williams in 1812. in building the Cantonment at the 11 mile Creek near Buffaloe, for a brigade Commanded by Brig. Genl. Smyth, of which the the Consolidated 5th & 13th. Commanded by me, composed a part, and that several other Officers have had similar Suits brought against them — The Circumstances attending the building that Cantonment were, as near as I can recollect, as follows; When Genl. Smyth abandoned the idea of invading Canada, he directed the Officers Command-ing Corps, to hut for the Winter — The Season being far advanced, and there being no eligible Site, or materials for building at or in the immediate vicinity of our

Colonel P. P. Schuyler to the Secretary of War on November 19, 1814, Page 1

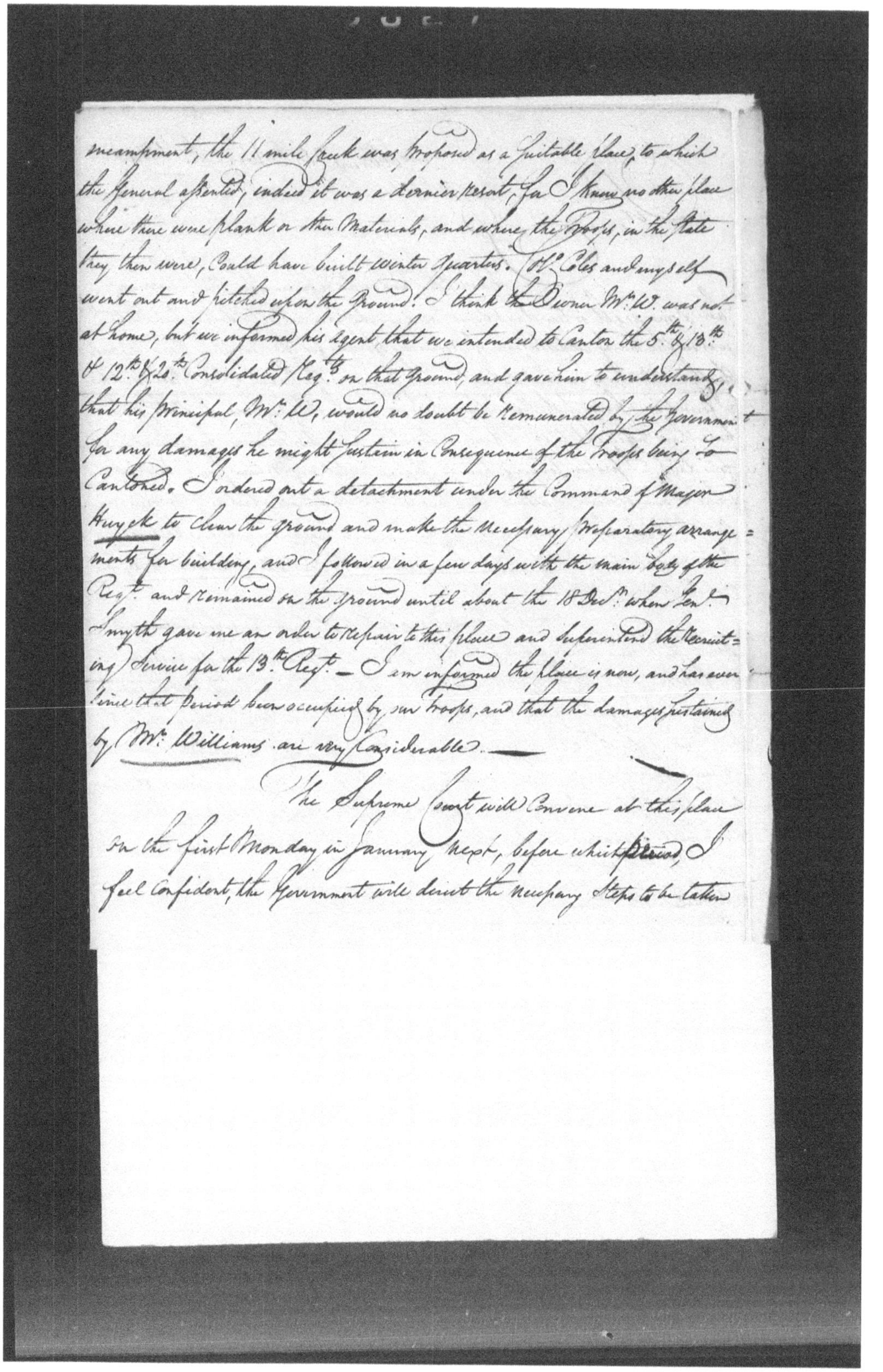
encampment, the 11 mile Creek was proposed as a suitable place, to which the General assented, indeed it was a dernier resort, for I know no other place where there were plank or other Materials, and where the Troops, in the state they then were, Could have built Winter Quarters. Col. Coles and myself went out and pitched upon the Ground. I think the Owner Mr. W. was not at home, but we informed his agent that we intended to Canton the 5th & 13th & 12th & 20th Consolidated Regts. on that Ground, and gave him to understand that his principal, Mr. W, would no doubt be remunerated by the Government for any damages he might sustain in Consequence of the Troops being so Cantoned. I ordered out a detachment under the Command of Major Huyck to clear the ground and make the necessary preparatory arrange=ments for building, and I followed in a few days with the main body of the Regt. and remained on the ground until about the 18 Decr. when Genl. Smyth gave me an order to repair to this place and superintend the Recruit=ing Service for the 13th Regt. — I am informed the place is now, and has ever since that period been occupied by our Troops, and that the damages sustained by Mr. Williams are very Considerable —

The Supreme Court will Convene at this place on the first Monday in January next, before which period, I feel Confident, the Government will direct the necessary steps to be taken

Colonel P. P. Schuyler to the Secretary of War on November 19, 1814, Page 2

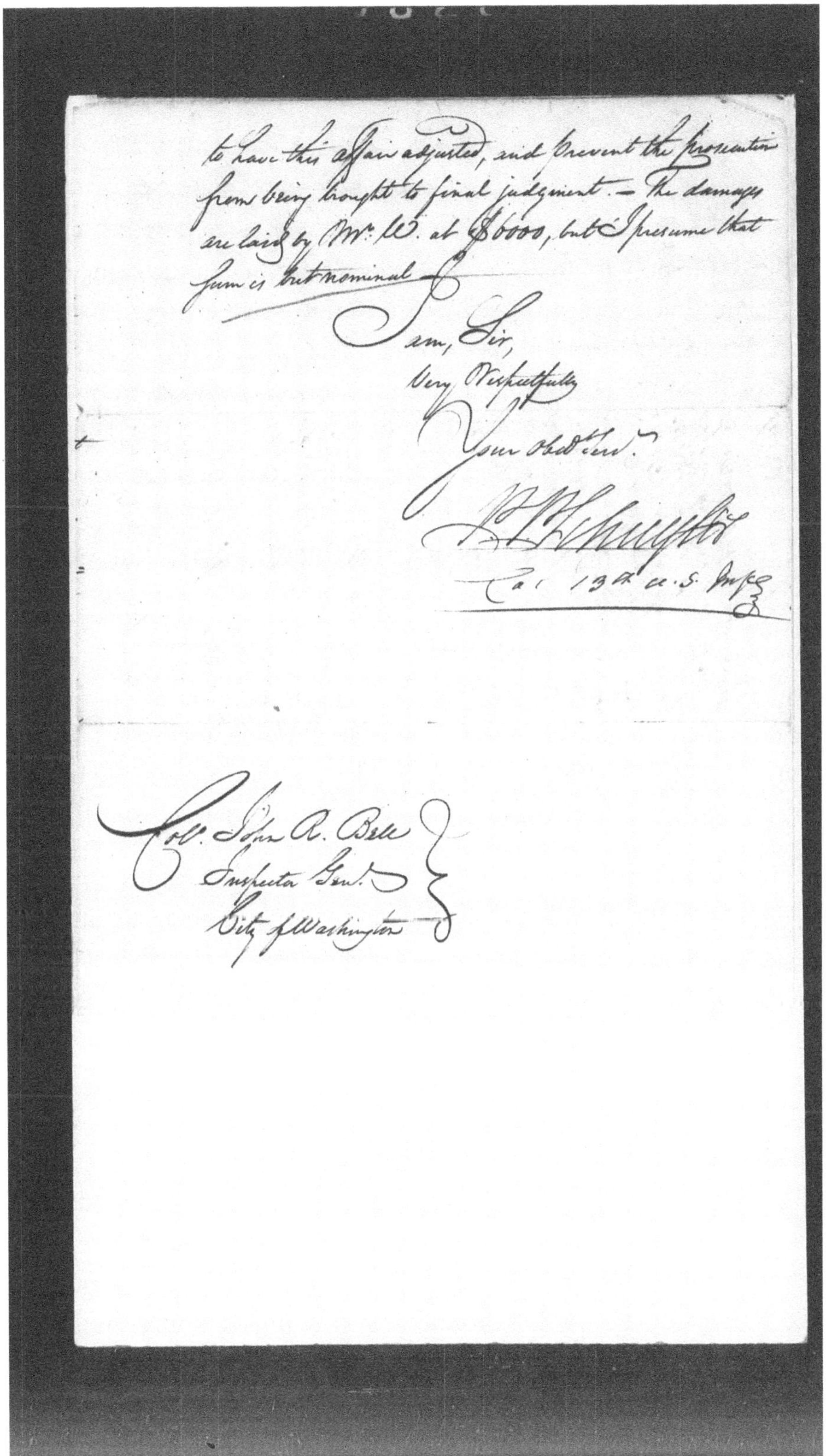

to have this affair adjusted, and prevent the prosecution
from being brought to final judgment. — The damages
are laid by Mr. W. at $6000, but I presume that
sum is but nominal —

I am, Sir,
Very Respectfully
Your Obed. Servt.
P. P. Schuyler
Col. 13th U.S. Infy

Col. John R. Bell
Inspector Genl.
City of Washington

Colonel P. P. Schuyler to the Secretary of War on November 19, 1814, Page 3

Colonel P. P. Schuyler to the Secretary of War on November 19, 1814
Transcribed Letter

Albany 19th November 1814

I have the honor to inform you, that said have some eight or ten days since , and would have reported sooner , but that I was in the daily expectation of receiving funds for the recruiting service , and further instructions on that subject . The funds in the hands of Capt. Having , who I found in charge of the recruiting service for the 13th Regt are, he states not more than sufficient to enlist five or six men, and should none be received in a few days, I propose leaving him in charge until my return from New York, where I am ordered to attend the General Court Martial which is to convene on the 15th Prop for the trial of Col. Coles . I will in the meantime procure and Communicate to your Office, all the information I can collect relative to the State and condition of the Reg.

I deem it my duty to state for the information of the Honorable Secretary of War, that a suit has been instituted against me for trespassing on the premises of a Mr. Jonas Williams in 1812 in building a cantonment at the 11 Mile Creek near Buffaloe, for a brigade Commanded by Brigadier General Smyth of which the Consolidated 5th 13th Commanded by me, comprised a part, and that several other Officers have had similar Suits brought against them.

The circumstances attending the building that Cantonment were, as near as I can recollect, as follows; when of Gen Smyth abandoned the idea of invading Canada, he directed the Officers Commanding Corps, to hut for the winter -- The season being far advanced , and there being no eligible site, or materials for building at or in the immediate vicinity of our

Next Page:

encampment, the 11 Mile Creek was proposed as a suitable place to which the General assented , indeed it was a dernier resort, for I know no other place where there were plank or other materials, and where the Troops, in the state they then were, could have built winter quarters. Col. Coles and myself went out and pitched upon the ground. I think the Owner Mr. W. was not at home, but we informed his agent , that we intended to Canton the 5th & 13th & 12th & 20th Consolidated Regt on that ground, and gave him to understand that his principal, Mr. W. would no doubt be remunerated by the Government for any damages he might sustain in consequence of the troops being to Cantoned. I ordered out a detachment under the command of Major Huyck to clear the ground and make the necessary preparatory arrangements for building, and I followed in a few days with the main body of the Regt and remained on the ground until about the 18th of December when General Smyth gave me an order to repair to this place and superintend the recruiting service for the 13th Regt - I am informed the place is now, and has ever since that period been occupied by our troops, and that the damages sustained by Mr. Williams are very considerable.

The Supreme Court will convene at this place on the first Monday in January next, before which period, I feel confident, the Government will direct the necessary steps to be taken

Next Page:

to have this affair adjusted and prevent the prosecution from being brought to final judgment. The damages are laid by Mr. W. at $ 6000, but I presume that sum is but nominal.

I am, Sir,

Very Respectfully Your Obedient Servant

Col. P. P. Schuyler

13th U.S. Infantry

Col. John R. Bell

Inspector General

City of Washington

The 1814 Land Sale and a Telling Clause

While these lawsuits and military disputes were unfolding, another development occurred behind the scenes. Jonas Williams received an offer from Juba Storrs & Company to purchase two parcels of his land, along with his sawmill, for the remarkable sum of fifteen thousand dollars. On June 5, 1814, Williams agreed to the sale but he added a crucial clause to the deed.

The deed stated that he was:

"saving and reserving any right of action"

against anyone in the United States Army for anything they had already done on the property.

In other words, even though he was selling the land, Williams ensured that he retained the right to pursue damages for the timber cut, the land cleared, and any other harm caused during the construction of the cantonment. These rights remained with him; they did not transfer to the new owners.

This small line in the deed reveals much. It shows that Williams fully intended to continue his legal fight, and it confirms that the military's use of his land was significant enough that he wanted those claims preserved even as he relinquished the property.

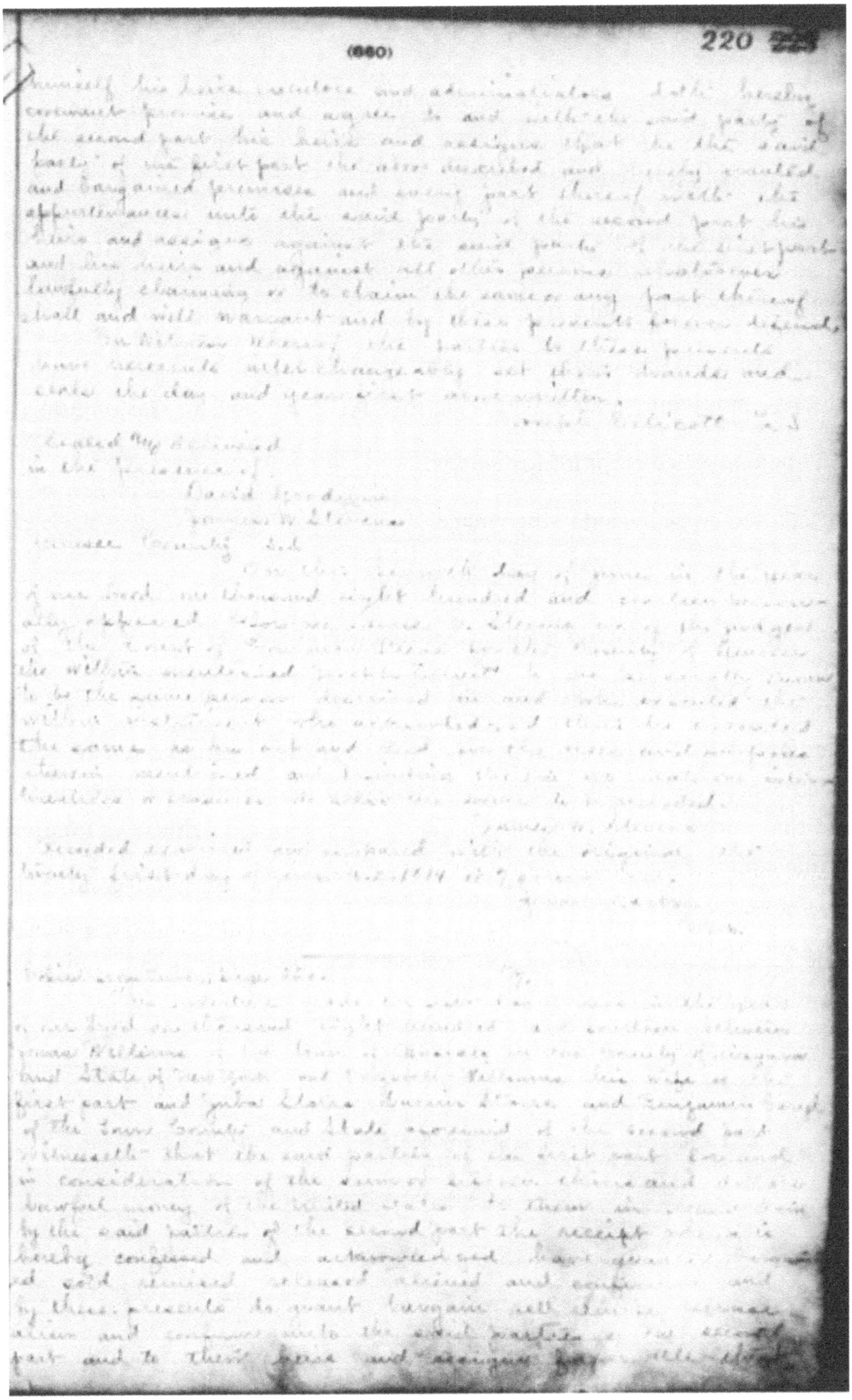
(880) 220

himself his heirs executors and administrators doth hereby covenant promise and agree to and with the said party of the second part his heirs and assigns that he the said party of the first part the above described and hereby granted and bargained premises and every part thereof with the appurtenances unto the said party of the second part his heirs and assigns against the said party of the first part and his heirs and against all other persons whatsoever lawfully claiming or to claim the same or any part thereof shall and will warrant and by these presents forever defend.

In Witness Whereof the parties to these presents have hereunto interchangeably set their hands and seals the day and year first above written.

Williams to and Juba Storrs Lucius Stores and Benjamin Caryl Deed Dated June 5, 1814, Liber 1, page 220.

A Forgotten Burial Ground

Fast-forward sixty-two years to 1876. By this time, the War of 1812 was already fading from living memory. The War Department began gathering information about where American soldiers from the conflict had been buried.

To compile this report, officials contacted Alex Sheldon, librarian of the Grosvenor Library in Buffalo. Sheldon, in turn, wrote to Timothy A. Hopkins, son of General Timothy S. Hopkins a War of 1812 veteran and longtime Amherst resident seeking his help.

Sheldon believed that Williamsville held one of the burial grounds used during the conflict, and he wanted accurate information before submitting his findings to Washington.

In his letter, Sheldon asked Hopkins for details:

• how many soldiers were buried in the village

• whether they were buried in coffins or mass graves

• whether any officers or enlisted men had identifiable graves

• and what legal status the burial ground held

Sheldon even noted that the government hoped to gather these scattered remains into a proper resting place perhaps even creating a site of beauty and remembrance though he added, with dry humor, that such a project might require "a very different House of Representatives" before funding could be secured.

This letter is significant because it shows that, decades later, federal officials still recognized Williamsville as a place where War of 1812 soldiers had been buried, and they turned to the Hopkins family for firsthand knowledge.

Copy of ms. in possession of Nelson S. Hopkins, Aug.3,1938.

93 Folsom Avenue
Buffalo Jary 15, 1876

Timothy A. Hopkins Esq

Dr Sir

The War Department has sent a request that I would give a report as soon as possible as regards the burial places of the Soldiers of the War of 1812. Some arrangement will doubtless be made to collect the scattered ones and raise memorials to them, and if there [are] any places where numbers are buried, they will be enclosed & taken care of as other National Cemeteries are.

I believe in your Village there is a spot of land which was devoted during the war for that purpose. Will you at your earliest convenience have the kindness to tell me about how many were buried in your Village; also how, whether in coffins or merely hurled into pits as in many places with their clothes on. Also are there scattered graves of Officers or Soldiers & if so can they be identified, and about how many are they?

Was the land given that they are buried in or by what tenure are the dead allowed there? Any information you can give me will be thankfully received and when I collect all that I can, I shall submit the letters with a report to the Department. I am of the opinion that if all of the dead Soldiers in & about your place can

Alex Sheldon Letter to T. A. Hopkins Page 1 Copy of Letter, housed at the Buffalo History Museum.

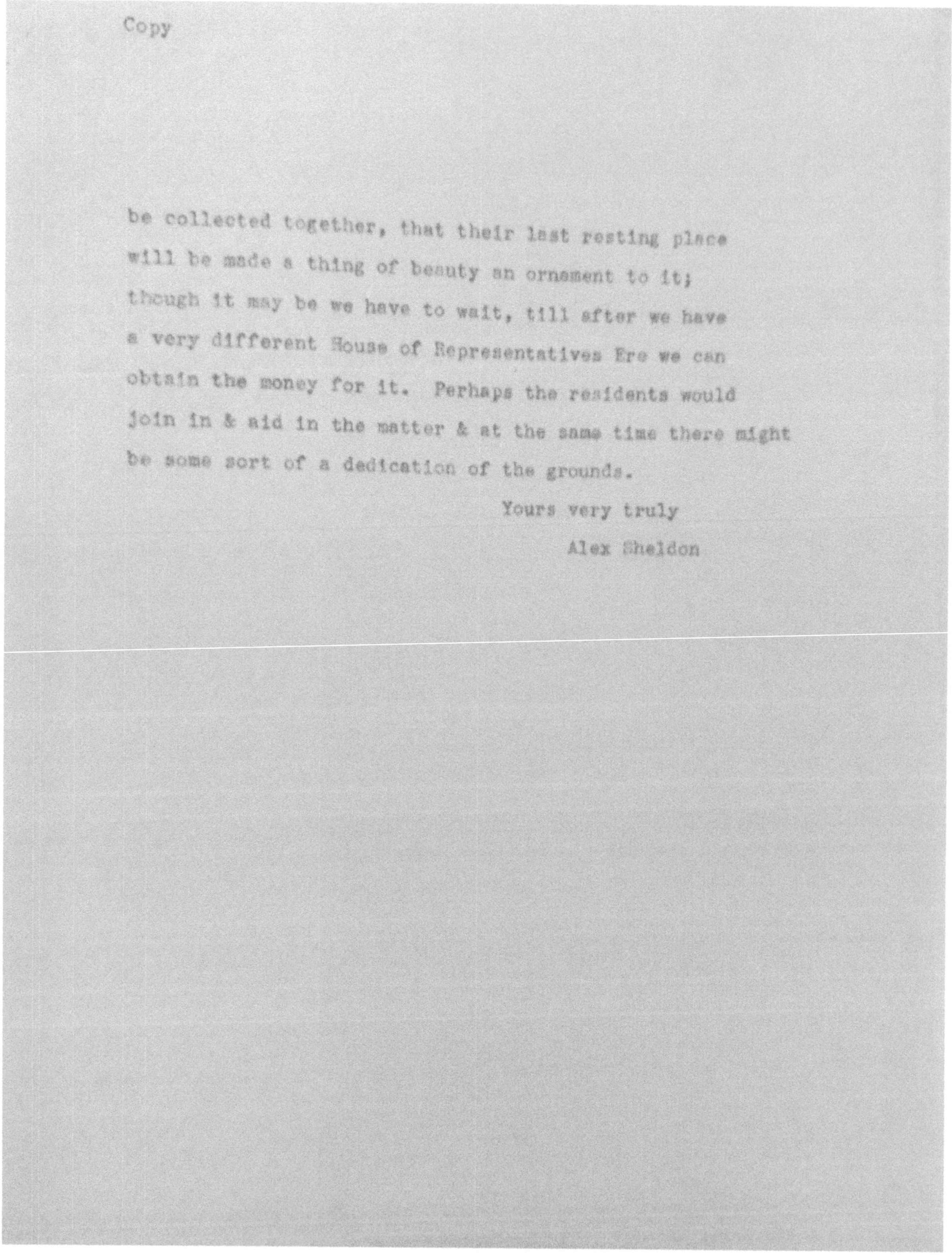

Copy

be collected together, that their last resting place
will be made a thing of beauty an ornament to it;
though it may be we have to wait, till after we have
a very different House of Representatives Ere we can
obtain the money for it. Perhaps the residents would
join in & aid in the matter & at the same time there might
be some sort of a dedication of the grounds.

Yours very truly

Alex Sheldon

Alex Sheldon Letter to T. A. Hopkins Page 2 Copy of Letter, housed at the Buffalo History Museum.

T. A. Hopkins Reply to Alex Sheldon in reply to his letter of January 15, 1876.

Just a few weeks after receiving that inquiry, I found Timothy Hopkins reply and that reply changed the course of my research!

On February 10th, 1876, Timothy A. Hopkins sent back a detailed response.

This letter is, without question, one of the most important pieces of evidence we possess regarding the village during the War of 1812.

Hopkins begins by setting the scene, reminding Sheldon that six thousand troops were ordered to Williamsville for winter quarters shortly after the declaration of war, and that many remained here until the end of hostilities.

But then, he provides a truly remarkable detail.

He explains that the main body of soldiers was stationed on the north side of Main Street. And then, almost casually, he drops a bombshell:

"...their burial place was in the rear of their quarters."

Wait. Their burial place?

We know about the Garrison Cemetery on Aero Drive. But a burial ground behind the Cantonment in the village?

Hopkins goes on to explain why we missed it. He notes that the land had been used for farming for more than forty years after the War.

By 1876, he writes, all visible traces of the graves had long since disappeared.

That single line "their burial place was in the rear of their quarters" is extraordinary. It identifies the location of a second War of 1812 burial ground, right in the village itself.

This letter doesn't just add a footnote to our research; it restores a piece of history that has been lost for generations.

Learning of the Cantonment Burial Ground is a breakthrough of the highest order.

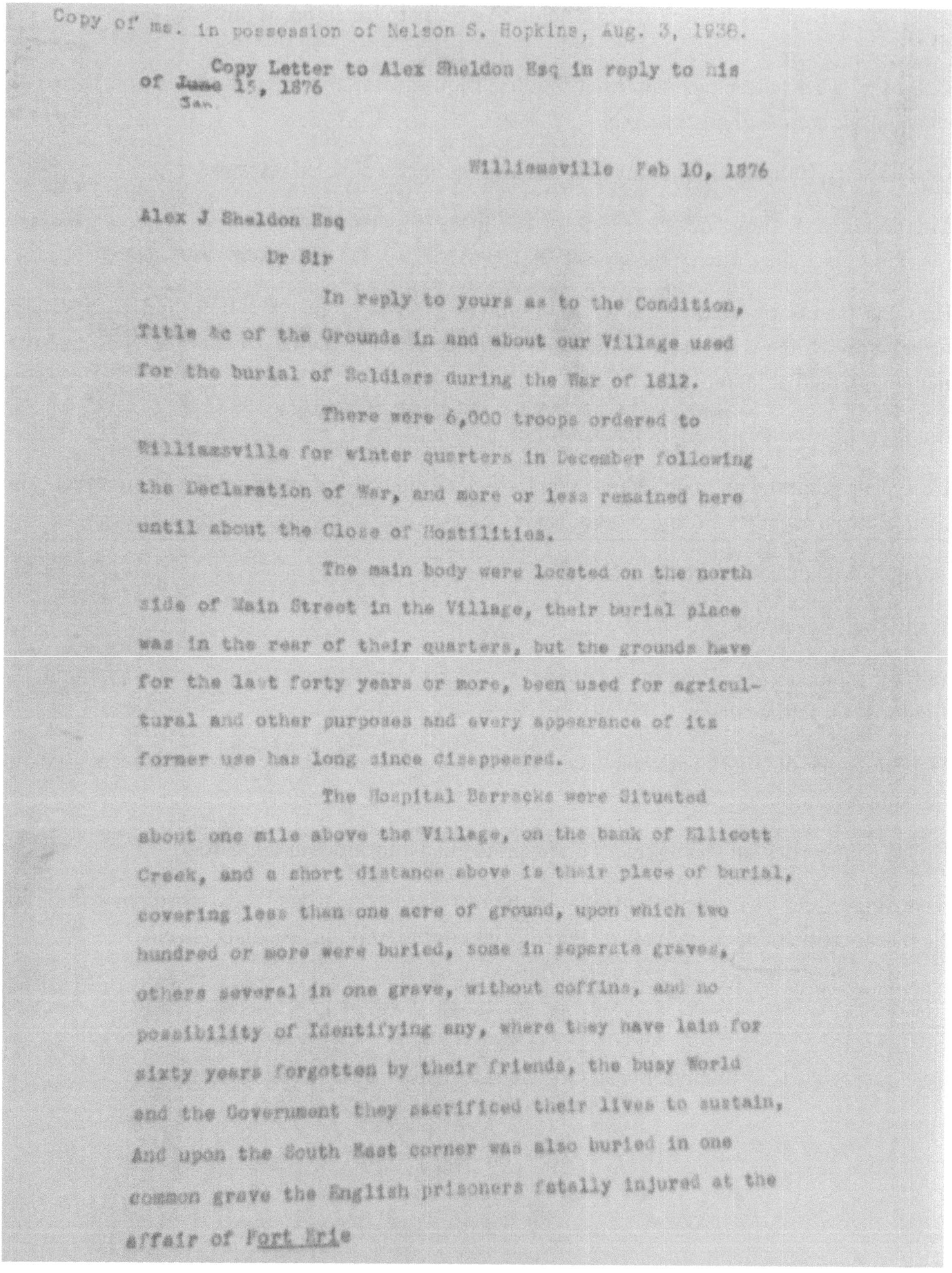

Copy of ms. in possession of Nelson S. Hopkins, Aug. 3, 1938.

Copy Letter to Alex Sheldon Esq in reply to his of ~~June~~ Jan 15, 1876

Williamsville Feb 10, 1876

Alex J Sheldon Esq

Dr Sir

In reply to yours as to the Condition, Title &c of the Grounds in and about our Village used for the burial of Soldiers during the War of 1812.

There were 6,000 troops ordered to Williamsville for winter quarters in December following the Declaration of War, and more or less remained here until about the Close of Hostilities.

The main body were located on the north side of Main Street in the Village, their burial place was in the rear of their quarters, but the grounds have for the last forty years or more, been used for agricultural and other purposes and every appearance of its former use has long since disappeared.

The Hospital Barracks were Situated about one mile above the Village, on the bank of Ellicott Creek, and a short distance above is their place of burial, covering less than one acre of ground, upon which two hundred or more were buried, some in seperate graves, others several in one grave, without coffins, and no possibility of Identifying any, where they have lain for sixty years forgotten by their friends, the busy World and the Government they sacrificed their lives to sustain, And upon the South East corner was also buried in one common grave the English prisoners fatally injured at the affair of Fort Erie

T. A. Hopkins Reply to Alex Sheldon in reply to his letter of January 15, 1876. Page 1 Copy of Letter, housed at the Buffalo History Museum

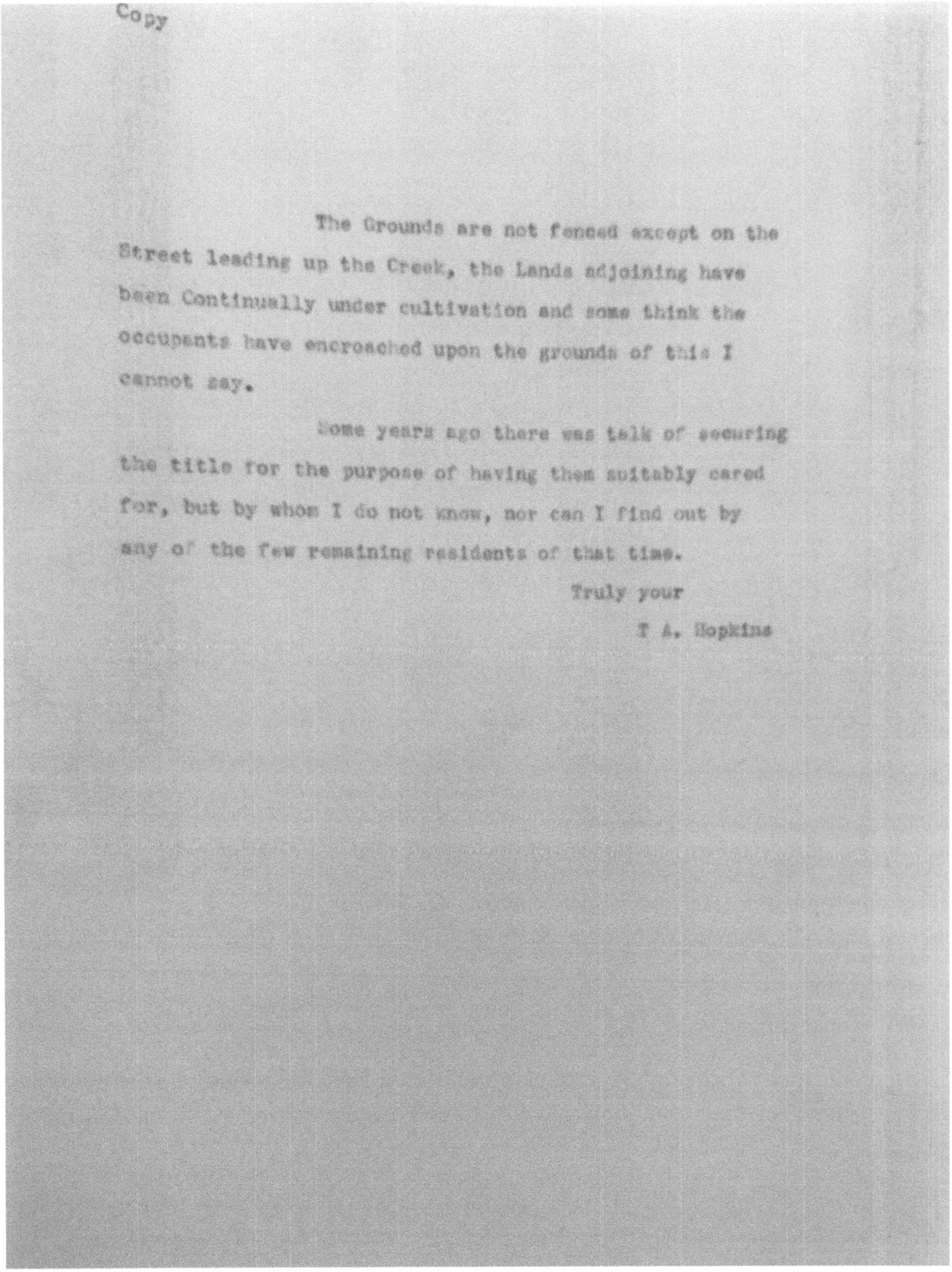

Copy

The Grounds are not fenced except on the Street leading up the Creek, the Lands adjoining have been Continually under cultivation and some think the occupants have encroached upon the grounds of this I cannot say.

Some years ago there was talk of securing the title for the purpose of having them suitably cared for, but by whom I do not know, nor can I find out by any of the few remaining residents of that time.

Truly your

T A. Hopkins

T. A. Hopkins Reply to Alex Sheldon in reply to his letter of January 15, 1876. Page 2 Letter, housed at the Buffalo History Museum

A Tragic Twist Of Fate

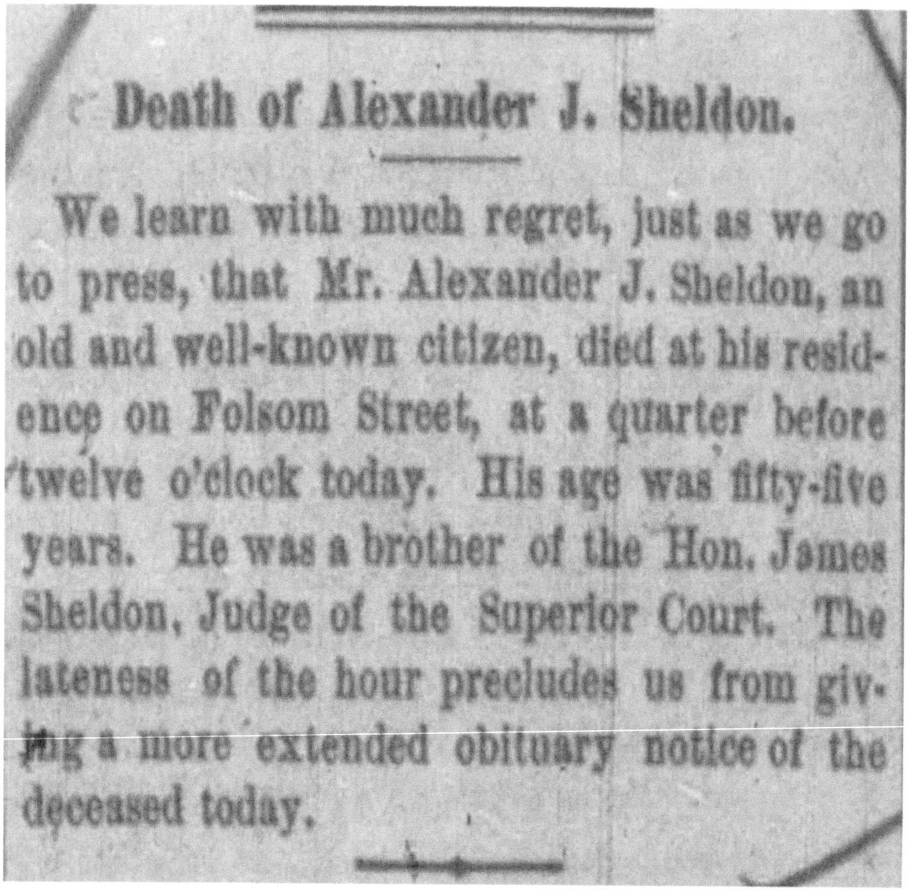
Death of Alexander J. Sheldon.

We learn with much regret, just as we go to press, that Mr. Alexander J. Sheldon, an old and well-known citizen, died at his residence on Folsom Street, at a quarter before twelve o'clock today. His age was fifty-five years. He was a brother of the Hon. James Sheldon, Judge of the Superior Court. The lateness of the hour precludes us from giving a more extended obituary notice of the deceased today.

The Buffalo Post Thursday March 2, 1876

But there is a tragic twist. This vital information never made it to Washington.

Tragically soon after receiving this letter, Alexander Sheldon died. The Buffalo Post reported that on Thursday, March 23, 1876, Mr. Sheldon passed away at his residence on Folsom Street.

With his death, the War Department's project for the War of 1812 Soldiers died too. The report was never filed.

The monuments at Williamsville were never built. And Hopkins' revelation about the burial ground behind the main cantonment was filed away in an archive, gathering dust for 150 years.

Now, I believe it's imperative that we finish the work Alexander Sheldon started: to identify this site and dedicate ourselves to restoring its sacred landscape for future generations.

Reconstructing the Landscape of 1812

To understand how this cantonment fit into the landscape, we must imagine Williamsville as it existed in 1812. The village was still in its earliest stages of development, with only a handful of structures, a sawmill, and a few scattered roads. The cantonment was not merely near the village it became one of the largest features on the landscape.

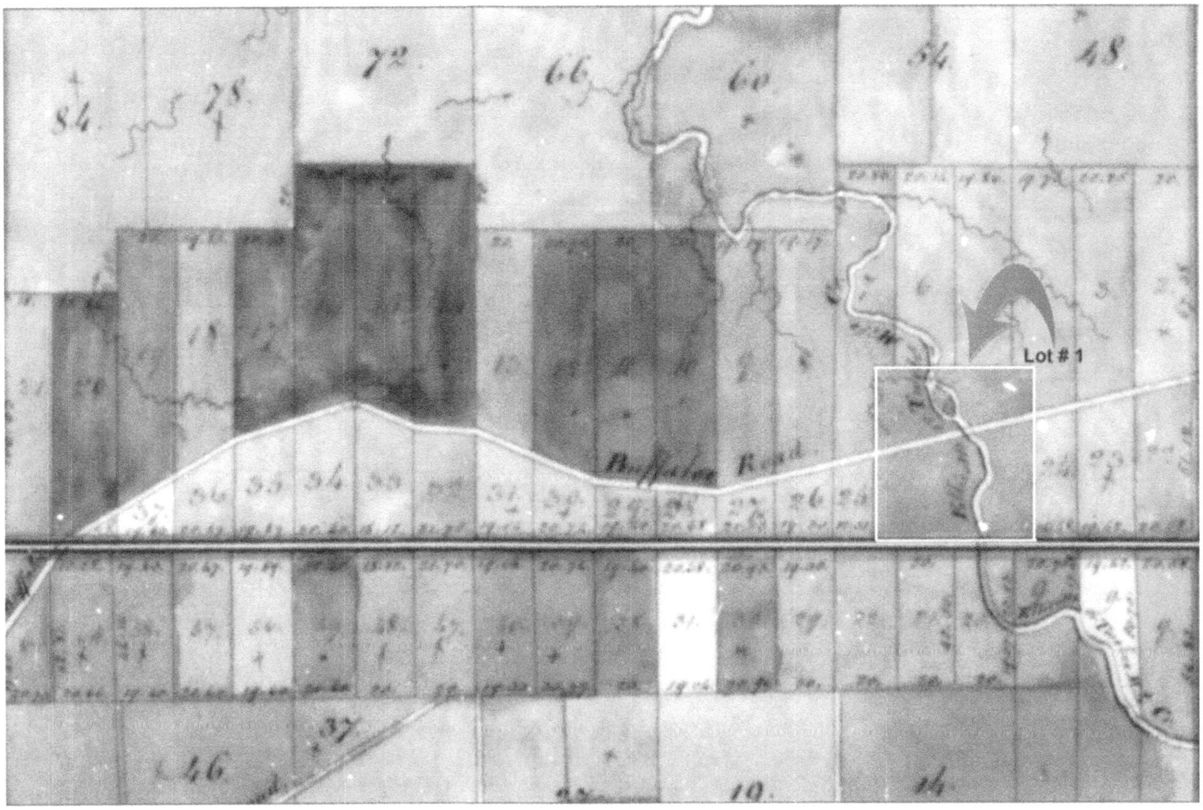

Holland Land Company Lot Map Showing Lot 1 Township 12 Range 7

Jonas Williams purchased Lot No. 1 in Township 12, Range 7 from the Holland Land Company a tract of roughly 300 acres. This single purchase would eventually become the Village of Williamsville.

Thus, when soldiers encamped here in 1812, they were occupying the earliest footprint of the village land still wilderness, still privately owned, and still decades away from becoming the community we know today.

Lot 1 Land Ownership in 1812

Before the war, Williams had already begun selling portions of his land, shaping the early footprint of the village. He sold roughly 135 acres east of Ellicott Creek to Isaac Bowman and conveyed nine individual village lots west of the creek along Main Street.

These transactions were not routine land deals they defined the layout of the settlement and help us interpret the military geography of Williamsville during the War of 1812: where troops could encamp, where barracks could be built, and how the cantonment fit into the emerging village landscape.

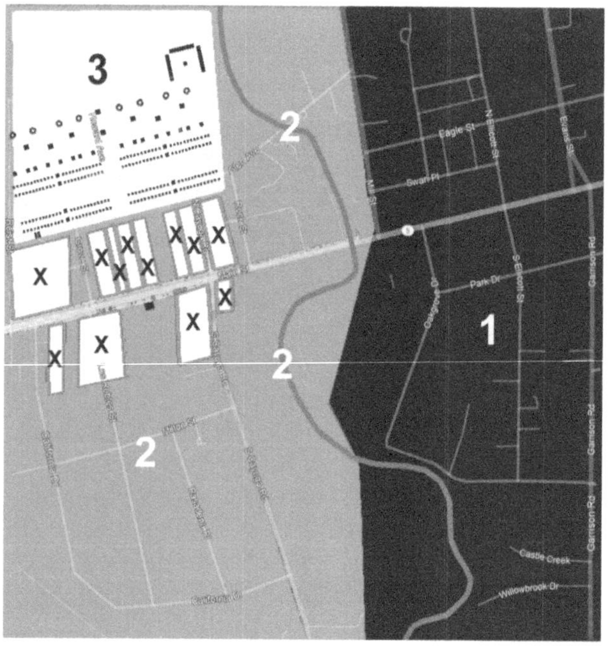

Lot 1 Land Ownership in the Year of 1812 Shown with present day Roads 1. Issac Bowman Land, 2. Jonas Williams land, 3. Cantonment and X. Individual Landowners

To understand where the Infantry Cantonment was established in December of 1812, we first need to picture what Williamsville looked like at that time.

These transactions weren't just routine land deals they played a major role in defining the layout of the settlement. And they help us interpret the military geography of Williamsville during the War of 1812: where troops could encamp, where barracks could be built, and how the cantonment fit into the emerging village landscape.

The map you see here reflects those early divisions, giving us a clearer sense of how the land was used and how the cantonment was nestled into a community still very much in its infancy.

Lot 1 Land Ownership in 1814

This map provides a valuable visual reference for one of the most important land transactions in early Williamsville history. In 1814, Williams sold a large portion of his property to Juba Storrs & Company, including the northern parcel that once contained the Infantry Cantonment.

The deed's clause preserving Williams's right to sue confirms the military's significant impact on the land.

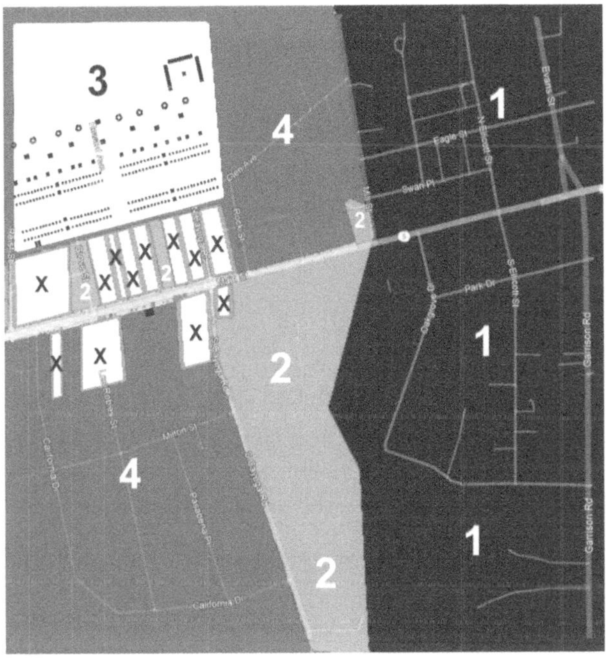

Lot 1 Land Ownership in the Year of 1814 Shown with present day Roads 1. Issac Bowman Land, 2. Jonas Williams land, 3. Cantonment, 4. Juba Store's Land and X. Individual Landowners

This map gives us a valuable visual reference for one of the most important land transactions in early Williamsville history.

In 1814, Jonas Williams sold a large portion of his property to Juba Storrs & Company, including the northern parcel, the very section that once contained the Infantry Cantonment.

By this point, the cantonment had already been built and occupied by U.S. troops, and Williams was still pursuing legal action over its use. But this sale didn't erase that history.

In fact, the deed included a clause that allowed Williams to retain his right to sue for damages, even after transferring ownership. So, when we look at this map, we're not just seeing land boundaries, we're seeing the footprint of a military encampment, a contested property, and a village in transition. This parcel helps us anchor the cantonment's location within the evolving landscape of Williamsville.

When we look at this 1854 village map, we're really stepping back into one of the earliest visual records of this community. The red outline marks the area where we believe the Infantry Cantonment once stood and what immediately stands out is just how open and undeveloped this land was at the time.

These parcels were held by Phebe Smith and Timothy Hopkins, and part of the outlined area belonged to T. A. Hopkins, the son of General T. S. Hopkins. So even in 1854, long after the cantonment itself had disappeared from the landscape, the property remained connected to families with deep local roots and strong military past.

The Cantonment Through Nineteenth-Century Maps

1854 Map of Williamsville

The red outline marks the area where the Infantry Cantonment once stood. In 1854, the land remained open and undeveloped. Parcels were held by Phebe Smith and Timothy Hopkins, with part belonging to T. A. Hopkins, son of General T. S. Hopkins. Even decades after the cantonment disappeared, the property remained tied to families with deep local roots.

Section of 1854 Map of Williamsville, New York

1866 Map of Williamsville

Section of 1866 Map of Williamsville, New York

The 1866 map shows the same area still vacant, with T. A. Hopkins listed as the only owner.

1880 Map of Williamsville

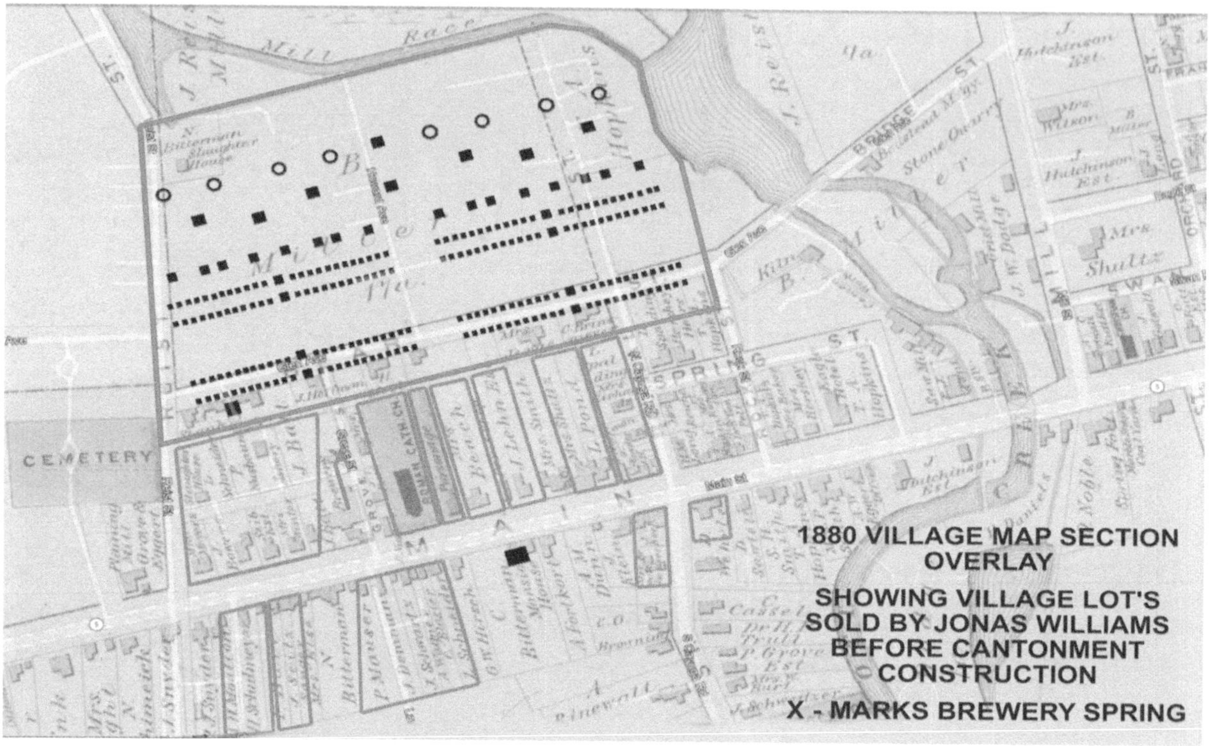

Section of 1880 Map of Williamsville, New York

As we move ahead to the 1880 village map, we can really see how the former Cantonment site continued to evolve or in this case, how it didn't evolve. The red outline marks the same area we've been following, and by 1880 the land is now owned by B. Miller and T. A. Hopkins.

What's interesting is that even after several decades, the property remains almost entirely vacant. The only real development we see is up in the northwest corner, where the N. Bitterman Slaughterhouse has been built. Otherwise, the landscape is still open, giving us a sense of how slowly this part of the village filled in compared to other areas.

It's another reminder of how these maps help us trace not just ownership, but the pace and pattern of growth and how the footprint of the old cantonment lingered long after the soldiers were gone.

1909 Map of Williamsville

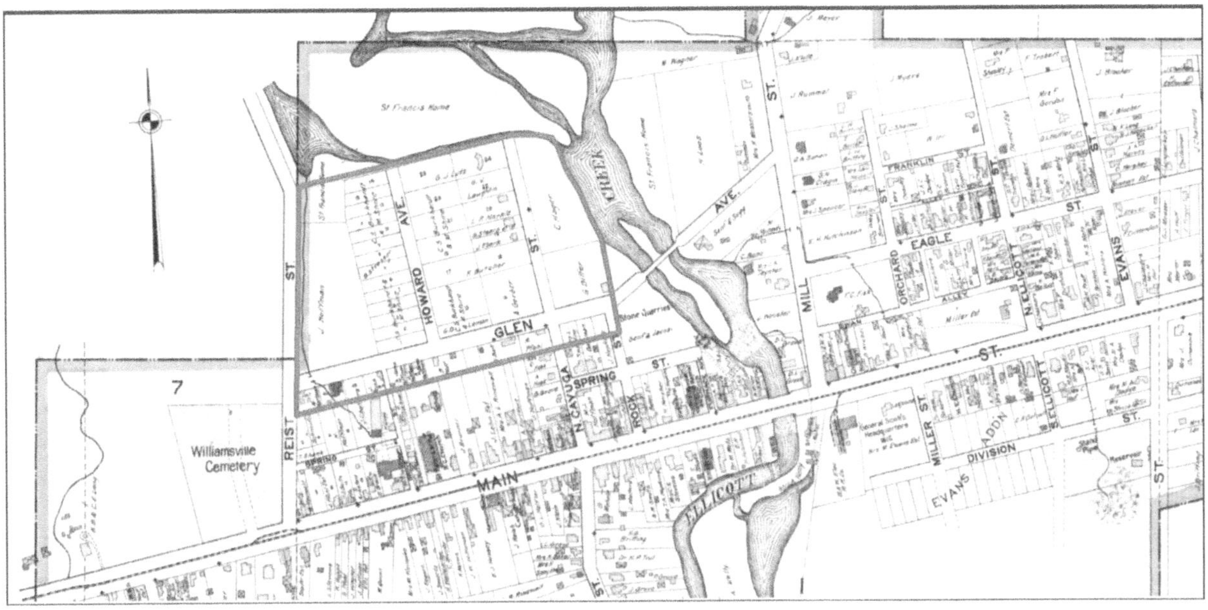

Section of 1909 Map of Williamsville, New York

Now as we move ahead to the 1909 village map, we finally start to see real change taking shape on the old Infantry Cantonment site. The red outline marks the same area we've been following, but unlike the earlier maps, this one shows the land fully divided into residential lots. By this point, five houses have been built on the property, marking the first significant wave of development on land that had remained open for decades. It's a striking contrast to the empty parcels we saw in 1854 and even in 1880.This map really captures the moment when the former cantonment area transitions from open ground with deep historical roots into a true neighborhood one that begins to resemble the village landscape we recognize today.

1927 Aerial Photo

1927 Aerial Photo of Cantonment Area

As we jump ahead to this 1927 aerial photograph, we can really see how dramatically the old Cantonment area has changed. The open land we saw on the earlier maps is now giving way to full-scale development.

By this point, the cantonment site has been almost entirely transformed, laid out, houses built, and the neighborhood begins to take on the shape we recognize today. It's a striking reminder of how quickly the landscape evolved once development finally reached this part of the village. This photo captures that moment when the last traces of the old military grounds disappear into the growing community, marking the end of one chapter of the land's history and the beginning of another.

LiDAR Imagery

LiDAR Imagery of the Cantonment Area

What we see here is a LiDAR image of the area. As you can tell, the entire area has been heavily developed over time.

But what is truly fascinating is that, even though the physical buildings of the cantonment have essentially disappeared, the land itself still holds a lingering mystery.

In his letter, Timothy Hopkins described a burial ground located directly behind the original cantonment.

Crucially, he also mentioned that the land that the burial ground was on had been used for farming for more than forty years.

But look just north of the Cantonment area, this LiDAR imagery strips away the modern surface to reveal something extraordinary:

You can still see the old plow furrows etched into the earth.We can't see these features on the surface today; they are hidden beneath the grass and soil.

But this image allows us to peer beneath the modern landscape and imagine where that forgotten piece of history might still rest. So, as we move forward, keep this in mind: layers of history don't always survive in our buildings or our roads. Sometimes, they survive in the archives, and in old documents, and if you know how to look in the landscape itself.

10
OTHER ACCOUNTS OF THE CANTONMENT WILLIAMS VILLE:

Mordecai Myers, Reminiscences, 1780-1814, including Incidents in the War of 1812

Major Mordecai Myers of the Thirteenth Infantry related this harrowing occurrence about sentry duty "I had in my company a young Irishman whose bravery was never doubted; but while I commanded the cantonment at Williamsville in the winter of 1812, a report was one morning current that O'Bryan had seen the Devil when on post at the sally port leading to the graveyard. Feeling that others might feel alarmed on going on the same post, I sent for O'Bryan and said to him, 'It is said that you saw the Devil last night when on post.' He turned very pale and said, 'I do not know whether it was the Devil, but something black, about the size of a horse without a head came rolling towards me. I challenged but received no answer; it was light, I plainly saw it still approaching; I sighted my piece to fire, but lost my strength and fell back against the cantonment where the relief soon after found me. 'I said, 'Are you afraid to go on the same post at the same hour tonight with me? Perhaps we may find that there was no Devil, but your own fear of spirits.' He said, 'Sir, I have never known fear, and I am willing to go with or without you - but I should prefer to have you go with me / I went and we remained on post two hours ;but the black Devil in the shape of a horse without a head did not renew his visit, and my having been on post with O'Bryan was known to all command, and it satisfied all that they might go on that post without meeting 'His Satanic Majesty'."

I believe the important part is "O'Bryan had seen the Devil when on post at the sally port leading to the graveyard."

This could substantiate the Hopkins letter talking about the Cantonment Burying ground behind the Cantonment,

Sue Miller Young, A History of the Town of Amherst, New York, 1818-1965 Amherst, N.Y. Town Board of Amherst, 1965

The barracks hospital was designated as a general military hospital in 1814, and soon 1,100 sick and wounded were brought from Lewiston. By the spring of that year, Williams Mills became an army base for more than 5,000 men assigned to the Niagara Frontier. Briefly, the settlement was considered to be the headquarters for the American forces during the war, with the arrivals of General Brown and General Winfield Scott, who made temporary quarters in the Evans house. During this time, soldiers were said to have been housed in log barracks erected along Ellicott Creek, north and parallel of Main Street in the vicinity of the present-day Garrison Road, which derives its name from these military structures. Historian Sue Miller Young puts the location of the arsenal and barracks on the north side of Main Street, between the site of the present-day SS Peter and Paul Church and Glen Falls. Military drills were held on the grounds near that church on Main Street.

Here Historian Sue Miller Young puts the location of the arsenal and barracks on the north side of Main Street, between the site of the present-day SS Peter and Paul Church and Glen Falls. Military drills were held on the grounds near that church on Main Street.

Ok so the Arsenal and Barracks as well as the Parade Ground are on the North side of Main Street and the Parade Ground is where SS Peter and Paul Church is today.

From A HISTORY OF BUFFALO AND NIAGARA FALLS

As the spring of 1814 advanced more troops were placed on the frontier. Their presence here brought a considerable amount of money for distribution among the people of this section, and a ready market at high prices was the consequence. Williamsville was the principal rendezvous for the troops, and a long row of barracks was built parallel with Main street, and a little north of that thoroughfare, while other buildings used for hospital and other purposes connected with military affairs were situated about a mile up Eleven-mile creek. By April 1 thirty or forty houses were built under contract, and business was in a flourishing condition. On April 10 Brigadier-General Winfield Scott, a fine, soldierly officer of about thirty years of age, arrived at Buffalo, and his superior officer. General Brown arrived soon afterwards.

This account also states: a long row of barracks was built parallel with Main Street, and a little north of that thoroughfare.

11

CONCLUSIONS REGARDING CANTONMENT WILLIAMS VILLE LOCATION

I conclude that the area bounded by Reist Street to the west, North Rock Street to the east, Glen Avenue to the south, and Amherst State Park to the north represents the most probable location of Cantonment Williams Ville and its associated arsenal. This conclusion emerges from the convergence of documentary evidence, eyewitness accounts, land use patterns, and surviving physical features within the landscape.

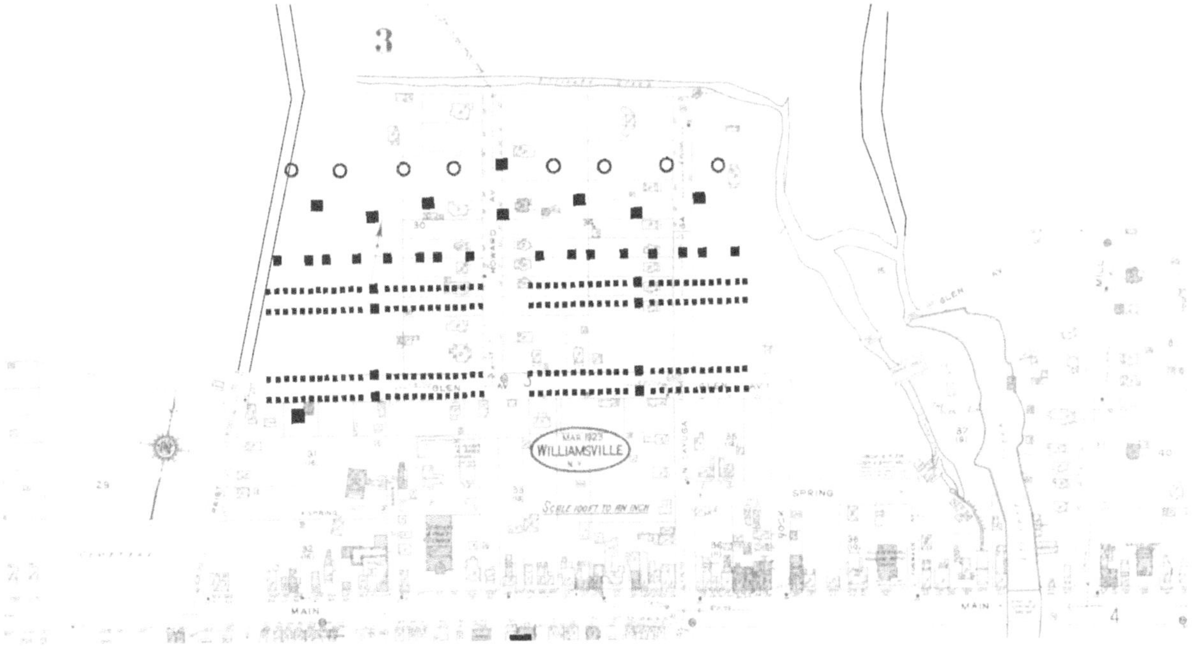

1923 Overlay of Proposed Cantonment Area.

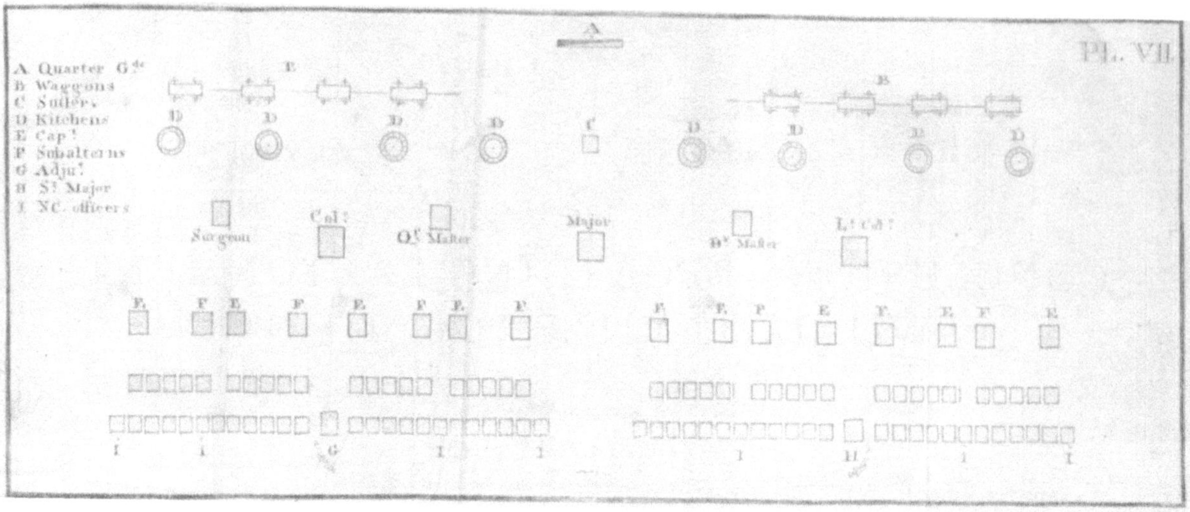

Von Steuben's plan for the order of encampment of a regiment making but two battalions 1802 Plate VII from, Regulations for the Order and Discipline of the Troops of the United States.

This area possesses several natural and logistical features consistent with an early nineteenth century military encampment. Most notably, it contains a natural spring capable of supplying a large body of troops with fresh drinking water. The spring head was located near the present day intersection of West Spring Street and Grove Street, aligning with early descriptions of Williamsville's reliable water sources one of the primary reasons General Smyth's officers selected the village for winter quarters.

Additional physical evidence strengthens this interpretation. Within the same bounded area, remnants of nineteenth century stonework survive ruins that I believe once formed part of the cantonment's infrastructure.

ADDRESS: East side of Grove Street (Tax parcel 5480 Main Street)	TAX ID: "O"	DATE OF CONSTRUCTION: Unknown; possibly 19th century	SIGNIFICANCE: D – Local Criterion C-1 Potential thematic group
LOT SIZE: 2.96 Acres	CONDITION: Deteriorated	NAME: Stone Wall Fragment	
		DESCRIPTION: This wall is located on the east side of Grove Street between Main Street/Rt. 5 and West Spring Street in the Village of Williamsville. The stacked stone wall has two distinct portions, the first is uncut with no coursing, and virtually no remaining mortar, the second is taller and uncut with irregular coursing and mortar. Possibly once part of a building foundation.	

19th Century Wall at Proposed Location

These structural remains, though fragmentary, correspond with the descriptions of long rows of log barracks and auxiliary buildings recorded in multiple accounts. Their presence within the proposed boundary reinforces the likelihood that this was the site of the principal encampment.

The conclusion is further supported by the early documents including the Hopkins manuscript, the Coles and Schuyler letters, and the 1814 land transactions—consistently place the main body of troops north of Main Street, extending toward Ellicott Creek and the lands later sold to Juba Storrs & Company.

The eyewitness testimony of Timothy A. Hopkins is especially valuable. He recalled that approximately six thousand troops were ordered to Williamsville for winter quarters and that their barracks extended from the village spring eastward toward the creek. His later correspondence with Alex Sheldon in 1876 also identified a burial ground "in the rear of their quarters," a detail that aligns geographically with the area under consideration.

Additional corroboration appears in *Other Accounts of the Cantonment Williams Ville.* Major Mordecai Myers's reminiscence describes a sentry post at the "sally port leading to the graveyard," reinforcing Hopkins's reference to a burial ground directly behind the cantonment. Myers wrote:

"O'Bryan had seen the Devil when on post at the sally port leading to the graveyard."

This vivid anecdote, though dramatic, provides a crucial spatial clue: the cantonment possessed a defined sally port and an adjacent burial area features consistent with a large, organized encampment such as the one proposed here.

Local historian Sue Miller Young further supports this interpretation. She placed the arsenal and barracks on the north side of Main Street, between the present-day SS Peter and Paul Church and Glen Falls, noting that military drills were held on the grounds near the church. This description situates the cantonment within the same northern corridor bounded by Reist, Cayuga, Glen, and the lands now comprising Amherst State Park.

Additional nineteenth-century accounts reinforce the same geography. A report from *A History of Buffalo and Niagara Falls* described "a long row of barracks…built parallel with Main Street, and a little north of that thoroughfare," precisely where the proposed cantonment boundary meets the elevated ground above Ellicott Creek.

Taken together, these accounts military, civilian, legal, and archaeological form a coherent picture. The area bounded by Reist Street, Rock Street, Glen Avenue, and Amherst State Park aligns with:

- the documented location of the Infantry Cantonment
- the presence of a natural spring
- the placement of barracks north of Main Street
- the proximity to Ellicott Creek
- the location of the burial ground described by Hopkins and Myers
- the nineteenth-century ruins still visible today
- the land parcels contested by Jonas Williams and later sold to Juba Storrs & Company

For these reasons, I conclude that this area represents the most historically accurate location of Cantonment Williams Ville and its associated arsenal.

12

THE CANTONMENT ARSENAL

I believe that the arsenal's powder room was located beneath the main structure, concealed within a hidden basement positioned at the rear of Sub-Lot 46 of Township 12, Range 7, Lot 1. On the Town of Amherst Lot Map, this parcel corresponds to 34 West Spring Street, situated directly between 87 Glen Avenue and 109 Glen Avenue.

During my broader research into War of 1812 arsenals across New York State, I encountered an illuminating article about the Syracuse Arsenal, which also possessed a concealed basement. The parallel was striking. In that case, the hidden chamber appears to have served as the storage area for black powder—an arrangement entirely consistent with early-nineteenth-century safety practices.

The article, written by Post-Standard columnist Dick Case, was titled:

Future Uncertain for Ruins of Stone Arsenal Built in Syracuse for War of 1812

1880s Photograph of the Onondaga Arsenal Off East Seneca Turnpike in Syracuse

Case described the discovery of the Syracuse arsenal's hidden basement in vivid terms:

"This retired U.S. Navy veteran and his son, Charlie, now a CBA high school senior, excavated the hidden basement of the aged leftover of what had been a three-story building" (Case, 2019).

This account reinforces my belief that the Williamsville Arsenal likewise incorporated a concealed subterranean chamber. Based on the physical evidence and the lot boundaries, I argue that this hidden basement lies off Glen Avenue, precisely between 87 and 109 Glen Avenue. At that location, two massive stone blocks cover a narrow opening that leads into what appears to be an artificial limestone chamber. The construction, placement, and materials strongly suggest that this was the arsenal's hidden powder room.

Photo of 87 Glen Ave. Location of Hidden Arsenal Basement

The two large stone blocks visible on the left side of the photograph mark the entrance to what I believe to be the concealed basement of the Williamsville Arsenal.

13

WILLIAMS VS. THE UNITED STATES

During the height of the War of 1812, the United States Army, acting under the exigencies of frontier defense, established a cantonment upon the property of Jonas Williams. In doing so, soldiers felled nearly all of Williams's finest timber, trees of notable quality and considerable value and occupied and damaged ground intended for the owner's private use and improvement. No formal contract was executed; instead, Williams agent received only verbal assurances that compensation would be provided at a later time.

Believing himself entitled to remuneration for the losses sustained, Williams initiated legal action. Colonels Coles and Schuyler were arrested for trespass, and a suit against the United States Army was instituted. Seeking a more definitive remedy, Williams enlisted the assistance of his friend, Judge Charles Townsend, to prepare a petition to Congress. The claim was first submitted during the Fifteenth Congress, Second Session, and wasn't referred to the Committee of Claims until the Sixteenth, Second Session, the body responsible for reviewing the multitude of petitions arising from wartime damage.

Then on 26 January 1821, the House of Representatives resolved itself into a Committee of the Whole to consider the committee's report. The recommendation was unequivocal: the claim, it declared, "ought not to be allowed." An effort was made on the floor to amend the resolution by striking the word *not*, thereby rendering the judgment favorable to the petitioners. When the question was put, however, the House declined to concur with the amendment. The original resolution was adopted, and the claim of Williams was formally denied.

The record offers no extended debate and no detailed justification for the decision. Nevertheless, the ruling accords with the prevailing sentiment of Congress in those years. The federal government, wary of establishing precedents that might open the treasury to a flood of similar petitions, held firmly to the principle that losses incurred during wartime whether by enemy action or by the necessities of military occupation did not, in themselves, constitute a debt of the United States. Unless a clear and formal agreement bound the government to pay, the Committee of Claims was inclined to reject petitions, however sympathetic the circumstances.

Thus, the claim of Jonas Williams, rooted in genuine injury and reflective of the hardships borne by frontier inhabitants, was set aside. His experience mirrors that of many whose lands and resources had been drawn into the nation's defense during the war: their sacrifices acknowledged in spirit, yet uncompensated in law. In this way, the story of the petition becomes part of the larger narrative of a young republic still defining the limits of its obligations to its citizens in times of national peril.

Yet the surviving evidence invites further reflection. In a letter to the Secretary of War, Colonel Schuyler wrote:

"The owner, Mr. W., was not at home, but we informed his agent that we intended to canton the 5th, 13th, 12th, and 20th Consolidated Regiments on that ground, and gave him to understand that his principal, Mr. W., would no doubt be remunerated by the Government for any damages he might sustain in consequence of the troops being so cantoned."

This assurance that Williams would "no doubt be remunerated" raises questions the official record does not answer. Had such statements been regarded as constituting a binding verbal agreement, might the outcome of the claim have been different? Were these letters overlooked, discounted, or deemed insufficient under the strict evidentiary standards of the Committee of Claims? Or, more pointedly, was Williams unjustly denied compensation that federal officers had led him to expect, or did those officers lack the authority to make such promises in the first place?

These questions cannot be resolved with certainty, yet they underscore the ambiguities that often-surrounded wartime obligations in the early republic. While Williams was away in Albany performing his duties as a legislator, and while his agent received assurances from military officers, no compensation ever materialized. Williams died in 1819, never knowing the fate of his petition, and his estate was left without redress for the losses he had sustained in service to the nation's defense.

Another lost chapter of Williamsville's early history now stands ready to be written into the historical record.

Jan. 26.] OF REPRESENTATIVES. 175

kins;" in which bills they ask the concurrence of this House. And then he withdrew.

The House resolved itself into a committee of the whole House on the bill from the Senate, entitled "An act for the relief of Morgan Brown;" and, after some time spent therein, Mr. Speaker resumed the chair, and Mr. Gross, of New York, reported that the committee had made some progress in the said bill, and desired him to ask leave to sit again.

Ordered, That the committee of the whole have leave to sit again on the said bill.

The House resolved itself into a committee of the whole on the report of the Committee of Claims on the petition of Charles Townsend and Jonas Williams; and, after some time spent therein, Mr. Speaker resumed the chair, and Mr. Hill reported the resolution submitted in the said report, with an amendment, to wit: strike out the word *not* before the word *allowed*, so as to make the resolution read, "Resolved, That the claim of Charles Townsend and Jonas Williams, of Buffaloe, in the state of New York, ought to be allowed."

And, on the question to concur with the committee of the whole in the said amendment,

It was determined in the negative.

The resolution submitted in the said report was then concurred in by the House, as follows:

Resolved, That the claim of Charles Townsend and Jonas Williams, of Buffaloe, in the state of New York, ought *not* to be allowed.

The House proceeded to consider the bill authorizing the Secretary of the Treasury to transfer certain balances of appropriations to the surplus fund; and the amendment reported thereto yesterday by the Committee of Ways and Means was read, and agreed to.

Mr. Meigs moved further to amend the said bill by striking out these words: "*Of the moneys appropriated for the payment of the interest, and redemption of the principal of the public debt, one million five hundred and ninety thousand seven hundred and forty-five dollars and seventy cents;*" so as to exempt this sum from being carried to the surplus fund.

And, the question being taken so to amend,

It was determined in the negative.

Ordered, That the said bill be engrossed, and read a third time to-morrow.

The House resolved itself into a committee of the whole on the bill to establish the district of Pearl river; and, after some time spent therein, Mr. Speaker resumed the chair, and Mr. McLane, of Delaware, reported the same without amendment.

Ordered, That the said bill be engrossed, and read a third time to-morrow.

The House resolved itself into a committee of the whole on the bill to regulate the entry of merchandise imported into the United States from any adjacent territory; and, after some time spent there-

January 26, 1821 Whole House Vote Rejected.

BIBLIOGRAPHY

Ashburn, Percy M. (1929). *A History of the Medical Department of the United States Army.* Houghton Mifflin, *448*, pp. 33-34.

Ashburn, Percy M. (1930). *American Army Hospitals of the Revolution and the War of 1812.* Bulletin of the Johns Hopkins Hospital, *46*, p.32.

Bingham, Robert W. Names of the United States Soldiers Buried in the Old Military Cemetery in Williamsville, New York Niagara Frontier Miscellany xxxiv, Buffalo Historical Society, 1947, p. 106.

Bingham, Robert W. *The History of Williamsville.* Niagara Frontier Miscellany xxxiv, Buffalo Historical Society, 1947, p. 97.

Case, D. (2019, March 22). Future Uncertain for Ruins of Stone Arsenal Build in Syracuse for War of 1812

Cruikshank, Capt. E. ed., (1897) The Documentary History of the Campaign on the Niagara
Frontier in 1814. Part II Published by Printed at the Tribune Office, Welland, Ontario, Canada

Devoy, John (1896). *A History of Buffalo and Niagara Falls,* The Buffalo Times.

Elliott , Charles Winslow (1937) Winfield Scott The Soldier and the Man

Geier, Orr and Reeves 2006, *Huts and History,* University Press of Florida.

Ketchum, William (1864). *An Authentic and Comprehensive History of Buffalo, with Some Account of Its Early Inhabitants Both Savage and Civilized* , pp. 256, 257

Litt, Williams & Whitehorne (1993) Death at Snake Hill Secrets from a War of 1812 Cemetery Dundurn Press

Livsey, Karen E. (1991). Western New York Land Transactions1804 – 1824, Genealogical Publishing Co.

Mann, James M.D.A.A.S, (1816) Medical Sketches of the Campaigns of 1812, 13, 14, Printed by H. Mann & Company

Myers, Mordecai (2010). *Reminiscences, 1780-1814, Including Incidents in the War of 1812,* Washington, 1900, Vol. 1, *Nabu Press.*

Pfeiffer, Susan & Williamson, Ron, (1991). *Snake Hill: An Investigation of a Military Cemetery from the War of 1812,* Dundurn Press, p. 40.

Shrauger, C., Sugg, R. & Troy, J. (1993). *Williamsville, New York: Where the Past is Present,* (3rd ed.), Williamsville Historical Society, p. 14.

Whitehorne, Joseph. (1992). *While Washington Burned The Battle for Fort Erie,* The Nautical & Aviation Publishing Company of America, pp.50, 55.

Young, Sue Miller. (1960). *History of the Town of Amherst,* Amherst Bee Article Series.

Young, Sue Miller. (1965). *A History of the Town of Amherst, New York: 1818-1965,* Town Board of Amherst, NY.

SOURCES

Artifacts Discovered at Garrison Park Digging Up the Past (2018) Amherst Bee November 07, 2018

Buffalo Gazette (Village of Buffalo, New York) on 10 January 1815

Buffalo Museum of History

Buffalo & Erie County Public Library

Checktowaga Historical Association

Documents of the Assembly of the State of New York, Vol. 22, 1917

Discover GIS Data NY

Erie County, New York

Fairchild Aerial Surveys 1927

F.W. Beers and Co. 1880 Village of Williamsville Map

Google Maps & Google Earth

Historic maps of Williamsville. 1854, 1880.

History of the Lehn Family. (1923), Amherst Bee, August 16, 1923.

Holland Land Company Maps

Township 12; Range 7; Willink's Strip; Buffalo Road; Ellicott Creek; Erie County (NY)

Township 11; Range 7; Buffalo Creek Reservation; Willink's Strip; Batavia Road; Ellicott Creek; Erie County (NY)

Source of original: Archives of the Holland Land Company, 1789-1869. Original project credit: SUNY Fredonia/GAA Amsterdam/NDR The Hague.

Maps. Retrieved from Archives and Special Collections at SUNY Fredonia

Letter dated December 12, 1812, Lt. Patrick McDonogh to his parents

Lineage Book by Daughters of the American Revolution

The U.S. National Archives and Records Administration

Niles' Weekly Register Article from January 10, 1815

Obituaries from the book Early settlers of New York State

Publications of the Buffalo Historical Society, Volume V

Valley Forge soldier's hut model by Phil Lord

War of 1812 Cemetery Timeline Courtesy of David Przepior

Williamsville Map From New century atlas of Erie County, New York. 1909. From government surveys, official records, and general compilations and notations by the company's corps of expert engineers and draughtsmen. Lew J.G. Orgden, Chief engineer, E.E. Whittemore, Asst. engineer, William Westgard, Chief draughtsman. Century Map Co. Publishing, Philadelphia, PA. Buffalo, NY. 1909.

Map. Retrieved from The Buffalo History Museum

SPECIAL THANKS TO:

Thank you to all the neighbors in the Lehn Spring development who generously opened their yards to me. I also want to extend my gratitude to everyone in the wider area who shared photographs and provided access to their property. I am especially grateful to:

- Amy Bartoszek and her son Matthew
- Brittany Jesserer
- Michele Didio
- Jennifer Windsor

ABOUT THE AUTHOR

J. A. Buscaglia's roots run deep in Western New York, he found himself drawn to the whispers of history echoing through its corners. It was this childhood fascination that ignited his lifelong passion for uncovering the tales of Western New York's pioneers and the tumultuous era of the War of 1812.

His expertise primarily lies in historical research, with a focus on the early settlers of Western New York, the Holland Land Company, and the events surrounding the War of 1812 on the Niagara Frontier.

Web Site:

https://jabuscaglia.com

ALSO BY J. A. BUSCAGLIA

Finding Jonas Williams: The search for Jonas Williams' Grave

In 1969 human bones were unearthed by workmen at a new office building in the Village of Williamsville New York and historians believed it was the unmarked grave of Jonas Williams founder of Williamsville, New York.

At the time members of the Amherst Historical Society and Town Historian, were willing to be certain that Williamsville had found its forbearer.

My extensive research will show that location is not the grave of Williamsville founder Jonas Williams and that his remains still laid undiscovered until now.

With there being such a diversity of opinion on the burial location of Jonas Williams and whether his remains have been discovered, I have decided to write my thoughts and observations on the topic. So, I am revealing all the information regarding this subject that I have been able to obtain from any source in order to determine the site, which I believe to be the location of Jonas Williams remains.

This is just my opinion from my historical research and my observations. It is not an academic paper but a first-person account of my findings.

Please be mindful that these are human remains and that the location I will reveal is on private property and trespassing is a violation of the law. I in no way endorse trespassing or breaking the law in any form.

Ancient Forts, Mounds, and Burial Grounds of WNY

Buscaglia ventures further back in time. Inspired by the writings of T. Apoleon Cheney and others, he embarks on a journey to unearth the remnants of a forgotten civilization. Drawn to the enigmatic earthworks and burial sites scattered throughout Western New York, Buscaglia seeks to preserve their legacy and unravel the mysteries they hold. With reverence for the sacredness of these sites, he emphasizes the importance of obtaining proper permission before exploring them, mindful of their significance to both the past and present.

The Garrison Burying Ground: A History of Cheektowaga's War of 1812 Cemetery

In this meticulously researched account, independent historian J. A. Buscaglia uncovers the forgotten legacy of the War of 1812 Cemetery on Aero Drive in Cheektowaga, New York. Drawing from archival deeds, military hospital records, and rare photographs, Buscaglia traces the evolution of this solemn site from its origins as a burial ground for soldiers who died of illness at the General Military Hospital near Williamsville, to its quiet transformation into a historical landmark.

The book explores the cemetery's complex land transactions, its ties to the Holland Land Company, and the efforts of local citizens like Col. John Bliss and John R. Evans to preserve the sanctity of the site.

Through aerial overlays, eyewitness accounts, and poignant reflections, Buscaglia brings to life the stories of over 300 American and British soldiers buried beneath unmarked soil.

With reverence and precision, The Garrison Burying Ground invites readers to rediscover a chapter of local and national history that has long been overlooked and to honor those whose names may be lost, but whose sacrifice endures.

Made in United States
Cleveland, OH
05 November 2026

47667156R00101